COST CONTAINMENT IN THE CLINICAL LABORATORY

Authors:

Clinical Chemistry:	Dee Evans, MT(ASCP) and Debra K. Frantz MBA, MT(ASCP)SH
Hematology:	Valerie L. Ng, Ph.D., M.D. Carol Hogan, MBA,MT(ASCP) and Eugene L. Gottfried, M.D.
Microbiology:	James I. Mangels, M.S.
Immunology:	Rhonda Emrich, M.S., and Ron Harbeck, Ph.D.

Editor and Publisher:
Edith Zak Helman, M.S.

A BERKELEY SCIENTIFIC PUBLICATION

International Standard Book Numbers: 910224-16-1

Lithographed In U.S.A.

Foreword

Since the advent of DRG's in the 1980's and the ever increasing role of managed healthcare in medicine, the clinical laboratory has become a cost center rather than the profit center it was in previous years. Therefore, the authors of Cost Containment in the Clinical Laboratory stress that in order to survive in this competitive environment and continue to provide quality services, the lab must control its costs by applying Quality Management Concepts to its operation, institute innovative and practical strategies to cut costs, and add computers and automation to its armamentarium.

In this book, experts in their laboratory disciplines, with business and laboratory management background provide specific cost control information which will help the laboratory prepare for the managed healthcare environment now and in the future.

Although Chapter I is written from a clinical chemistry point of view, many of the business principles can be valuable to managers of other sections of the laboratory.

In Chapter II, the authors discuss cost-effective strategies for the clinical hematology laboratory. Topics include: Why Was This Test Ordered?, STAT and Priority Testing, and New and Uncommon Tests.

The author of Chapter III discusses three factors that impact on cost savings in the Clinical Microbiology Laboratory: factors related to patient testing (utilization issues, what tests should be ordered, and how and why to change ordering practices), technical operation factors (quality collection, storage, transport, and processing of specimens), and factors dealing with labor management operations.

In Chapter IV, the authors discuss the role that automation, instrumentation, assay kit configuration, selection of appropriate tests, the design of the laboratory, and specific test modifications have in controlling costs in the immunology laboratory.

Edith Zak Helman, M.S.
Publisher

Table Of Contents

Chaper II - Clinical Hematology

Chapter III - Clinical Microbiology

Chapter IV - Clinical Immunology

Chapter I

Cost Containment in The Clinical Chemistry Laboratory

Dee Evans, MT(ASCP)
and
Debra K. Frantz, MBA, MT(ASCP)SH,

Stanford University Hospital Laboratories,
Stanford, CA
and
Kaiser Permanente Medical Center Laboratories,
South San Francisco, CA

I. INTRODUCTION

WHY IS COST CONTAINMENT IMPORTANT?

Historically, well-managed laboratories reigned as one of the hospital's profit centers, easily covering their costs and contributing toward the overhead of the institution and the departments which lost money. Competent laboratory managers and supervisors recognized the importance of controlling costs within their departments. By carefully spending their money, they were able to stay within their budget and maintain the laboratory's contribution margin or profit.

Further, the advent of multichannel chemistry analyzers allowed chemistry laboratories to provide a greater number of results with a faster turnaround time and minimal increase in cost. Technologic advances provided physicians with the ability to quickly and easily look at a patient's profile and to assess the functions of several different physiologic systems at once, rather than having to wait for individual test results.

In recent years, however, the cost of health care has spiraled out of control. Because of increasing costs, profit centers within hospitals were forced to increase their prices in an attempt to improve the contribution to cost centers which lost money. A number of solutions have been developed in an attempt to curb this costly cycle. Below are listed several cost containment models that have been introduced over the last decade, and their impact on laboratory services.

DRG'S

As most readers of this book undoubtedly know, in the early 1980's Medicare introduced a new plan for reimbursement--Diagnosis Related Groups, or DRG's. The premise behind DRG's was that for a specific diagnosis, a plan for the treatment of the patient could be established, and the steps to accomplish successful treatment of the patient could be quantified on a dollar basis. Medicare began paying hospitals based on DRG's, and for the first time, hospitals were challenged with caring for a patient within a cost limit established by an outside payer. If patient care exceeded the dollar amount established for the DRG, the hospital had to absorb it; if patient care cost less than the dollar amount, the hospital could keep the excess, thereby realizing a net gain. As a result of payment plans such as Medicare's DRG's, laboratories were pressured by hospital administration to contain their costs of testing, as well as to temper price increases.

CAPITATION

In capitated health care plans, carriers team with employers to enroll thousands of members in their health plan. The carriers pay health care providers a negotiated amount per month per "covered life" to care for the health of their members. This shared risk model gives providers incentive to minimize utilization of laboratory (as well as other) services.

In a capitated environment, the laboratory faces greater pressure to achieve the lowest cost possible for a test. This could even mean sending the test to a reference laboratory that can perform the analysis for a lower cost. The laboratory carefully analyzes the cost of every test and determines if it is cheaper to send the test out or perform it on site. These make-versus-buy decisions take center stage in the capitated model.

THE HEALTH SYSTEM

Strategic advantage as well as cost containment are driving forces in the development of the health system model, where vertical integration allows a continuum of care to be offered as one package.[1] In the health system, institutions such as hospitals, clinics, group practices, skilled nursing facilities, home health care agencies, etc. join together to provide 'cradle-to-grave' service in the healthcare market. Frequently, health systems develop an in-system reference laboratory, a model in which the systems' facilities send their routine work to their own (usually off-site) reference laboratory and keep only the most necessary stat testing on-site. The advantages to such a reference laboratory are the economies of scale that can be achieved by centralized testing, such as maintaining full capacity on high volume instruments and enjoying the purchasing power gained from buying materials in high volume.

THE ROLE OF QUALITY MANAGEMENT IN COST CONTAINMENT

The cost effectiveness of having a strong central laboratory as opposed to fragmented laboratory work done at many different sites throughout the institution is obvious, in that there is common utilization of resources needed to provide laboratory services. Everything from the cost of glassware and glassware washing equipment to the salary costs of the Laboratory Director and other management and support personnel is reduced when there is no duplication of these costs in every separate laboratory. In our experience, the best ways a laboratory can promote a strong central laboratory as opposed to fragmentation to many sites is the development of the philosophy of a customer-oriented laboratory and the commitment of energy and resources to meet that goal.

All of the quality management tools which have become so popular in recent years, from "Quality Circles" to "Service Excellence" to "Total Quality Management" to "Process Management and Improvement" have in common the recognition that the laboratory must be committed to providing continually improving service to its customers, just as private business organizations have done.

These management concepts emphasize that the laboratory has many customers including the patient, physicians, nurses, and other hospital personnel and departments, as well as clinics and outside medical groups. They promote learning what our customers want from the laboratory by the use of customer analysis surveys and increased communication.

Total quality management (TQM) groups, which consist of representatives from different divisions of the organization, have been very successful in bringing people together in a spirit of cooperation with a common goal. There are often simple solutions to eliminating barriers to providing good service when people from different departments understand the problems and needs of other areas. Almost without exception, these simple solutions that allow laboratories to provide improved service also result in lower costs.

Process management involves defining baselines and benchmarks to determine where the laboratory stands, operationally and financially. Indicators used to measure levels of service and cycle times are developed. Groups from different areas are appointed to work on ways of improving these services. Once the improvements have been made, there are periodic checks to ensure that the gains that have been made are continuing.

CASE STUDY IN TOTAL QUALITY MANAGEMENT: MORNING ROUNDS FOR PHLEBOTOMISTS

Our laboratory has worked hard at improving turnaround times, since lengthy turnaround time was the most common complaint from our customers. As a major TQM project, we identified improving turnaround times for the morning draw. Physicians, especially those who had private practices and who visited their hospital patients before going to their offices, needed to have the test results from the morning draw available no later than 8:00 A.M. This included all priorities (e.g., "stat", "expedite", "routine") of the common monitoring tests such as electrolyte and renal panels.

The steps that were taken to improve this service involved bringing in a substantial number of phlebotomy personnel at 5 A.M. to begin drawing blood. The phlebotomy personnel worked with supervisors to determine the most efficient way to get the job done. Phlebotomy teams were created and runners collected the specimens periodically and brought them to the lab. Eventually a pneumatic tube was installed to facilitate transport of specimens.

Some chemistry personnel were scheduled to arrive at 6 A.M. rather than 7 A.M. and they worked closely with the technologists from the night shift to get all of the monitoring tests from the morning draws done as quickly as possible. Calibrations and maintenance were done either before 6 A.M. or after 9 A.M.

This effort was extremely successful. Calls to the lab from physicians and units to change priorities from "routine" or "expedite" to "stat" diminished markedly. The perception of the lab as a customer-oriented service improved dramatically, as demonstrated by customer interviews.

Quality assurance issues such as this example are directly related to cost, in that the time involved in supervisor's responding to physician complaints had previously been considerable. Physician requests for change of priority from "routine" or "expedite" to "stat" on their morning draw patients were previously common, and responding to each of these requests required time to locate the sample, change the priority in the computer, and repour the sample if it had been aliquoted but could not be found. Now, more than two years later, we continue to monitor our performance and survey our customers to ensure that they are satisfied with our service.

IMPROVING THE BOTTOM LINE

A very simplified Income Statement for a not-for-profit organization is shown below[2]

Revenues
<u>-Variable expenses</u>
Contribution to Fixed Expenses (Contribution Margin)

Simple math shows that to increase the contribution margin, revenues must increase, variable expenses must decrease, or a combination of the two must occur. Even in complex organizations whose income statements are far more complicated than the one in this illustration, these basic truths still apply.

TYPES OF COST

As in the laboratory world, the accounting world has its own vocabulary and jargon inherent to their business. Any good accounting text book[3] can provide definitions of unfamiliar terms; accounting terms that will be used frequently here are defined below.

Fixed Costs

Costs that are unchanged as volume changes within the relevant range of activity. Examples of fixed costs within a laboratory include the costs of management, supervisory and clerical personnel, proficiency testing, and licensing fees.

Variable Costs

Direct material and direct labor costs that change as volume changes within the relevant range of activity. Examples of variable costs include the cost of testing personnel, phlebotomists, lab personnel who process the specimens, and cost of reagents and supplies.

Direct Costs

A synonym for variable costs.

Indirect Costs

All manufacturing costs other than direct material and direct labor, including overhead. Examples of indirect costs include the cost of reference testing, the laboratory information system (LIS), the personnel involved with the LIS, and utilities for the physical plant.

These terms are not necessarily discrete; some are used interchangeably and some costs fit into more than one category. Also, there is not necessarily an absolutely correct category for every cost; some costs fall into gray areas or might fit appropriately into opposing cost categories (for example, an argument could be made on both sides that the cost of a lease for a piece of testing equipment could be considered indirect or direct). When in doubt about how to classify a cost component, seek the opinion of the institution's financial manager.

II. IDENTIFYING AND UNDERSTANDING THE COMPONENTS OF COST

There are three components that comprise cost: materials, labor, and overhead. In this section we will examine each component in detail as it relates to the clinical chemistry laboratory.

Be aware that under most circumstances, "equipment acquisition" is not an exercise limited to acquisition of a chemistry analyzer only. Especially in terms of cost, reagents, disposables, controls, standards, the number of tests performed to arrive at a result (e.g., testing in duplicate, etc.), and proficiency testing must be carefully considered. These are all components of a *test system*, and we will use this term throughout the text.

MATERIAL COSTS

Features and Operational Characteristics

The chemistry laboratory manager or supervisor faces interesting challenges in trying to balance purchasing decisions. Cost effectiveness is always a prime consideration, but not at the expense of accurate and precise results. Frequently, the manager must consider the wishes of a clinical chemist or pathologist, who may have an interest in a particular instrument.

While certain instruments may meet or exceed technical requirements, they may also exceed the capital budget, or have high recurring costs. Clearly, the cheapest machine is not always the best choice for the laboratory; many other factors must be considered. In this section, we attempt to describe all of the factors which must be considered, both technical and economic. The manager who has carefully considered all factors can make the most thoughtful purchasing decision.

Test Volume: Consider the features that are important for your lab's needs in terms of the projected volume of tests to be done, *vis-a-vis* other instruments in place in your lab. Will the new instrument be primarily a stat instrument or one used primarily for routine profiles? Will the instrument be operational 24 hours a day, 7 days a week? Manufacturers of chemistry analyzers typically offer different products specifically designed to meet volume needs; be sure that you understand which of their offerings is appropriate for your volume and operation. Also, future growth may be an important consideration; if your laboratory's volume is projected to increase or decrease, your instrument needs may change significantly.

Sample Volume: Some instruments are better equipped to handle smaller sample sizes than others. The manager must consider the importance of this feature, e.g. does the laboratory service a pediatric unit or other unit where sample volume is important? Again, consider future growth--is your facility adding a bone marrow transplant unit, where small sample volume may be crucial?

LIS Interface: Will the instrument be interfaced to a laboratory computer? If so, then it is important to decide whether a unidirectional or bi-directional interface is required or desired, and that depends on the capabilities of both the instrument and your LIS.

In a bi-directional interface the LIS tells the instrument what tests are to be done on a given sample and then the instrument tells the LIS what the results of those tests are. Instruments which have a unidirectional interface require that an operator enter the ordered tests, but the instrument can then transmit the results to the laboratory computer after verification by the technologist. As you will see in the next section, computer interfacing also can have a significant impact on the cost of the project.

Barcoding/Primary Tube Sampling: Does the instrument have the capability to read bar codes for specimen identification? Does your laboratory have a need for primary tube sampling, or will pouring specimens continue to suffice? Primary tube sampling eliminates most of the intralaboratory errors in sample identification and speeds work, but it requires a larger sample volume and may not be an appropriate solution for a small chemistry laboratory.

Wet versus Dry Reagents: The chemistry manager must decide which type of reagent is most appropriate for his or her laboratory. In a wet reagent system, tubing is hooked up to reagent reservoirs and automatic pipettors. Most of the problems which occur with wet instrumentation involve a break in the integrity of the systems used to transport liquids, either reagent or sample, or a loss of calibration in devices used to measure fluid volumes. Troubleshooting by lab personnel can be tiresome and time-consuming, but is usually successful. A worn plunger on a calibrated syringe can be replaced, tubing can be changed, a leaky connection can be repaired.

In a dry reagent system, the manufacturer packages reagents in such a way that liquid reagent is no longer measured by the instrument. Thus, problems previously experienced with calibration error or mechanical breakdown of pipetting devices are eliminated, and convenience is greatly improved. Frequently, however, dry reagents are part of a closed system that eliminates flexibility in reagent and/or methodology choice and allows the manufacturer to set pricing higher.

Space Considerations: An important factor may be the space that you have available for the instrument. Do you have enough room for a floor model, or is space a limiting factor in your laboratory? Reagent storage and the length of time before reagents expire can be a major consideration; having to recalibrate if there is a change in lot number or shipping date to comply with Clinical Laboratory Improvement Amendments (CLIA) makes this critical, especially if the time required for calibration or calibration verification is long and is disruptive to the efficient functioning of your department. Can calibration or calibration verification be done on the midnight shift or during some other slow period, or are you forced to work it into your busy day? Do you have a back-up instrument that you can use while you calibrate, or is your laboratory down during this period?

Troubleshooting/Instrument Maintenance: The revolution in microchip circuitry and incorporation of microprocessors into modern instrumentation have radically changed the troubleshooting process with both wet and dry instruments. It is now imperative to have a telephone available near the instrument so that you can check electronic readings while you are talking to someone on the manufacturer's "hot line". In some cases instruments are connected by modems so that electronic checks can be done directly by the service center.

Sometimes adjustments in settings on the instrument will solve the problem, but often a new circuit board or a step motor that is controlled by a circuit board must be installed, necessitating an express shipment from a central service location which results in a prolonged downtime. Some companies have approached this problem by offering a kit that has a set of duplicate boards, but these are usually expensive. It is extremely important that you get the company's commitment to response time in writing when negotiating the instrument acquisition. It is also helpful to deal with companies who market primarily to hospital laboratories.

Several years ago, our chemistry department purchased an atomic absorption instrument that had been used mainly in industrial applications. When we needed service, we were told that the repairman could come in about a week and a half! Only by saying "people will die" were we able to get the company's attention. Although one of the authors (Evans) had been involved in the selection of the instrument, no manufacturer of atomic absorption instruments was willing to give a firm commitment about response time. In retrospect, it might have been prudent to look at other methodologies or at outsourcing those tests which could only be done by atomic absorption.

Manufacturer's Experience Curve: Buying a newly-marketed instrument is courageous but usually not advisable. There is a certain degree of something...maybe notoriety...in having serial #1 but then the lab enjoys the privilege of uncovering the bugs, weaknesses, and errors in design that are solved and redesigned during the first year. Additionally, there are unknown interferences that become known after a time; even if your laboratory is getting the instrument at no charge, it may not be cost effective. The same holds true for software that is to be used with the instrument; you may want to wait for the bugs to be worked out before getting a newly released software upgrade.

On the other hand, you don't want to buy an instrument at the end of its life cycle, just before its replacement is released. When you are no longer able to get replacement parts, unless they are used ones that have been scavenged from old instruments at the company's service center, then you will know you have waited too long.

Literature Search/Experience with the Instrument: Be sure to read all the articles and reviews you can about any instrument that you may be considering, but remember that research sponsored by the company that manufactured the instrument may not be totally objective and unbiased. Try to get as much exposure to the instrument as you can at trade shows, conventions, and professional seminars. Most companies will be happy to arrange an in-house demonstration of their instrument and should provide a list of reference hospitals where the instruments are in use.

Supervisors who have had the instrument for awhile are probably your best source for learning both the good and bad features of the instrument. A site visit to a hospital laboratory that is comparable to yours in size and operational characteristics is a wise investment. Try to make the arrangements on your own and not as the guest of a sales representative. You are more likely to get a candid assessment of the instruments performance if the company representative is not present.

Costs of Capital

Once the laboratory supervisor, manager, clinical chemist and pathologist have agreed on the features and operational characteristics required of a test system in their specific operation, attention can be directed toward the cost of the ideal instrument, and the cost/benefit trade-off of each characteristic. In this section, we describe the major considerations pertaining strictly to capital costs.

A major expense for any chemistry laboratory is the cost of capital. Before making a capital equipment acquisition, the astute manager or supervisor thinks carefully about what capital costs may be involved with the acquisition, since not all of the costs involved are as straightforward as the purchase price of the instrument. Below are listed the capital costs that must be considered.

Equipment Cost: This is the actual cost for the instrument(s) that the laboratory is purchasing, and any related equipment required to operate the instrument, such as an uninterrupted power supply (UPS), or a personal computer.

Sales Tax: Usually the sales tax assessed on the acquisition of equipment is capitalized by the Accounting Department. Be sure to check with them before you have agreed to an acquisition price with your vendor; sales tax can be a significant portion of the sale on some of the larger, more expensive chemistry analyzers.

Shipping: Again, the Accounting Department will usually capitalize any shipping charges. Check with your accountant before negotiating a shipping cost.

Insurance: Handled the same as shipping, as far as the accountants are concerned. Check with yours.

Remodeling/Capital Improvements: In acquiring a chemistry analyzer, be sure to discuss with the manufacturer any facility changes which are required for the new instrument. For example, some chemistry analyzers need a deionized water source; how will your laboratory ensure a constant supply of deionized water? Will the new instrument fit through the existing door of the laboratory, or will the doorway need to be altered? Is there sufficient space in the laboratory for the instrument, or will cabinetry have to be rearranged to accommodate it? If the instrument is a countertop model, is there enough clearance between the equipment and overhead cabinetry? Will reagent storage for the new instrument require a new refrigerator or freezer?

Make sure your vendor has explained thoroughly any safety issues involved with the new instrument. Does the new instrument allow the laboratory to introduce a new procedure that requires a new chemical fume hood (or changes to an existing chemical fume hood)? Installation of a chemical fume hood is an expensive capital cost that should be considered, if necessary. It is wise to involve your Facilities Engineer before the acquisition of new equipment.

Does the new instrument generate enough heat that it may affect the environment? Even a small piece of equipment may create enough heat that the laboratory's heating/air conditioning/ventilation system may be overtaxed, and certainly an additional refrigerator or freezer will affect the laboratory environment. Work with your Facilities Department to ensure that your laboratory's environment can accommodate new equipment.

Manufacturers recommend dedicated electrical lines for most instruments. Again, your Facilities Department must understand the specifications required for your new purchase; the cost of installing a dedicated electrical source is part of the capital acquisition cost. Also ensure that there are enough electrical outlets in the area where the instrument is to be installed for additional peripheral equipment such as computer workstations and printers. If a modem hook-up is required for troubleshooting, make sure there is an extra phone jack in the area.

LIS/MIS Costs: It is also necessary to involve the LIS manager. As often as possible, chemistry instrumentation should be interfaced to a laboratory's LIS, and for greatest efficiency, a laboratory should take full advantage of the interface and other computer capabilities that an instrument offers. Most LIS vendors have a set fee for interfacing a new instrument, but the cost is usually significant and should be discussed early in the acquisition process. The equipment vendor, the LIS vendor, the LIS manager, and the chemistry manager should discuss the system expectations for the new instrument. As discussed in the previous section, issues such as bi-directional versus unidirectional interfacing, barcoding, and primary tube sampling should be considered, as well as the additional capital costs involved with these improvements. If the instrument will be placed in a remote location, costs for networking or telecommunications must be considered.

Acquisition Method for Capital Equipment

In addition to paying thorough attention to capital costs, the acquisition method is important to the final cost of the instrument. There are three common acquisition methods in use in the laboratory industry.

Direct Purchase: This is the method which usually results in the lowest price for the instrument, since the vendor receives full payment immediately. Advantages to direct purchase are that the institution owns the instrument outright, and can depreciate the asset as they would any other piece of capital. A laboratory that owns an instrument should expect to pay a lower price for reagents and other consumables used with the instrument. On open systems, where different brands of reagents can be used, the laboratory is free to try different reagents and is in a position to negotiate the lowest price from competing vendors. Additionally, at the end of the useful life of the instrument, most manufacturers will accept it as a trade-in for acquisition of new equipment, or a laboratory can try to sell the old instrument.

There are disadvantages to direct purchase that a laboratory must consider. For an expensive piece of equipment, the amount of purchase must be budgeted sometimes up to a year in advance, which means thoughtful planning 12-18 months before the actual purchase is made. "Trading up" to a later model instrument or newer technology may be more difficult if the laboratory owns the instrument; provisions for acquiring new equipment may be a part of the pre-purchase negotiation between the laboratory and the vendor. If new technology from another vendor becomes available before the instrument is fully depreciated (rarely fewer than five years, and sometimes seven years), the institution may be reluctant to consider funding the newer equipment.

Lease: When lump sum capital dollars are not available, a chemistry laboratory can choose to lease an instrument. In a lease arrangement, monthly payments are spread over a specified period of time (60 months, for example) and are based on the purchase price plus interest and *executory costs* (i.e., taxes, insurance, and maintenance). While in most cases the lease is capitalized, leasing can still represent an advantage over purchase when cash availability is an issue for the institution.

Other advantages that leasing offers[4] are 100% financing at fixed rates, no money down and protection for the laboratory against inflation and increases in the cost of money.

In a leasing arrangement, the laboratory is protected against obsolescence if the vendor allows upgrades to the leased equipment. There may be other advantages from an accounting standpoint, so it is important to engage your institution's Accounting or Finance Department when considering a lease. They can help you better understand your institution's accounting for specifics in a leasing arrangement, such as buy-out arrangements (bargain purchase versus fair market value, for example) at the end of a lease, and help you make an educated decision with regard to what works best for the laboratory and the institution.

The major disadvantage of a lease is that of locking a laboratory into a long-term relationship with one vendor. If new technology becomes available from a competing vendor, a laboratory must still fulfill their leasing obligation with their original vendor. If the vendor's leasing company (manufacturers rarely finance leasing themselves) does not provide for early termination, the chemistry manager may want to negotiate this point up front or consult with their own institution's Accounting Department. It is possible that the laboratory's organization has a company through which they could lease the instrument under more advantageous conditions.

Another disadvantage of leasing is higher reagent and consumable prices, in most cases. If the vendor is sharing in the risk in leasing, he will generally increase the pricing of his products. This is another reason why it may be wise to lease the instrument through your own institution, as opposed to the leasing company that the vendor recommends.

Reagent Rental: In a reagent rental agreement, the cost of the instrument is included in the reagent cost charged by the vendor. The laboratory signs an agreement which usually commits it to a minimum volume of reagent purchased over a specified period of time at a set price.

The advantage to a reagent rental is that there are no capital costs for the laboratory. Whereas direct purchase requires a large amount of money up front, and leasing requires capital money spread over time, reagent rental requires no capital money. If a laboratory finds itself in need of equipment and there is simply no money in either the capital budget or contingency funds, reagent rental can be quite attractive. Additionally, since reagent rental locks a laboratory into one brand of reagent, manufacturers are generally agreeable to updating older instruments as new ones become available.

Reagent rental is generally the most costly option for equipment acquisition. Expenses which could otherwise be capitalized are folded into operating expenses, driving the cost per billable test up.

Reagents

Most chemistry managers recognize that recurring expenses, such as those associated with reagents, are where manufacturers make their biggest margin. Thus, manufacturers have a strong incentive to keep reagent pricing as high as possible.

Open systems allow the laboratorian to use different brands of reagents on the instrument, thus permitting price competition between manufacturers. Open systems offer the advantages of using different brands of reagents and the possibility of developing new tests on the analyzer. Because of these two factors, price competition is more likely to exist on an open system. Open systems, however, are inclined to be more complicated to use, requiring more training and in some cases, identification of "key operators" who become the resident experts on the system. Further, if the manufacturer of an open system also manufactures reagents, they may put pressure on the laboratory to use their reagents or else face difficulties in getting prompt service.

Closed systems, which require that a laboratory use reagents that are specifically designed for use in a particular instrument, lock a laboratory into one brand of reagent. In closed systems, reagent delivery is usually accomplished through a mechanical device (such as a cartridge, a slide, or a reagent pack) that is patented. The patent effectively prevents another manufacturer from encroaching on the test system; thus the laboratory has no option in reagent selection, methodology selection, or in reagent pricing. Closed systems, while potentially costlier than open systems, often offer convenience and ease of use over open systems. When offered as a random access system, they are frequently useful in busy laboratories with a heavy stat load and a high number of operators, since their use requires minimal training and user attention. Their primary disadvantage, as mentioned before, is the requirement that the laboratory use the reagent system specifically designed for the instrument, thereby limiting the flexibility of the instrument. Both systems offer advantages and disadvantages to the user; again, the manager must carefully weigh the costs against the features and characteristics required in the chemistry laboratory.

Part of any negotiation should be a discussion about service--response time from the vendor for service calls, loaner equipment if necessary, etc. Though the practice has somewhat decreased over the years, some manufacturers will hold the laboratory over the proverbial barrel when service is necessary, if the laboratory uses another brand of reagents.

Even in a cost-effective test system, reagents represent the biggest component of material costs over the life of the instrument. Because of this, in making an acquisition decision, a chemistry manager should always request a five-year cost estimate for reagents, and recurring reagent cost should be a major factor in the decision. A low instrument price, or a "give-away" instrument is no bargain if the recurring reagent costs are high. It is also wise to negotiate a limit on annual price increases from manufacturers; some manufacturers implement price increases as often as twice a year. Most often, these price increases are part of the manufacturer's annual marketing plan. The frequency and amount can be discussed as part of the acquisition agreement. Sometimes a vendor will offer free goods instead of a reagent price reduction; if this is the case, ensure that the free goods are something the laboratory would purchase anyway.

Sales representatives are often given "samples" of reagent by their companies; the expense of such samples are also a part of marketing expenses for which the manufacturer plans. Ask your vendor representative if he or she can provide free product during the year. No matter what agreement is reached with the vendor, *put it in writing*. If the vendor representative promises free product over a certain period of time, a written document protects the laboratory if the vendor rep leaves the company.

Disposables

To a lesser extent, disposables that are instrument-specific can be negotiated as part of the acquisition agreement. Instrument-specific disposables are usually not as costly as reagents, and many disposables are not instrument-specific but are the type of disposable that might be used anywhere in the laboratory (such as pipet tips, etc.). However, if the selected instrument has costly disposables, their purchase should be handled in the same way as reagents: ask for a five-year cost quote and control price increases.

Controls, Standards, and Proficiency Testing

Controls and Standards: Controls and standards can be a costly component of a test system, so it is important to carefully consider their use. There are guidelines in CLIA[5] under Subpart K, *Quality Control for Tests of Moderate or High Complexity, or Both* which govern the use of calibrators. These guidelines require that calibrators:

1. be run at least as frequently as the instrument manufacturer's instructions specify;
2. be run at least every six months or whenever the following occur:
   a. A complete change of reagents for a procedure is introduced, unless the lab has been able to demonstrate that there is not a change in the calibration when a new lot number of reagents is used and quality control values are not affected by a change in reagent lot numbers;
   b. There is major preventive maintenance or replacement of parts that may influence test performance;
   c. Quality control values reflect a trend, a shift, or are outside of their accepted limits;
   d. Whenever the laboratory's established schedule for verifying the reportable range requires more frequent calibration.

The CLIA guideline for the use of controls states that control material must be run at least as frequently as the instrument manufacturer's instructions specify, and at least two levels of control material must be run each day of testing. Control usage is one of the contributors to the cost of a test that can be very expensive for laboratories.

It is very important to use control ranges that are realistic and are related to the ranges of clinical relevance of the test. Some of the newer automated instruments are capable of achieving precision that is much narrower than those required for clinical relevance. When control ranges are set to the very narrowest ranges that are possible on such instruments, and then a small shift occurs, the amount of time spent troubleshooting and money that is wasted can be enormous. This involves the cost of repeat controls that must be run when a test is even slightly out of control for the second time in a row.

Technical time spent on the documentation of out-of-control situations can be considerable. Supervisory time spent on checking to ensure that the correct procedures were followed and that all the documentation, troubleshooting, and calibration required were done whenever an out-of-control situation occurred can border on the ridiculous. Thus, the setting of realistic and clinically relevant ranges is one of the major considerations in running a cost-effective chemistry department.

Recalibration must now be done according to the CLIA'88 guidelines given above. With some instruments, recalibration is a major time-consuming procedure and can mean that the instrument being recalibrated may be "down" for a significant period of time. Again, it is important to emphasize the necessity of setting realistic quality control ranges to avoid calibration when possible.

Proficiency Testing Material: Part of the cost of the test system is the cost of proficiency testing (PT). This is an especially important consideration when a chemistry laboratory is looking at implementing a new test. Often, a chemistry laboratory receives appropriate PT material as part of a series of PT to which they already subscribe. If the laboratory can make use of their existing PT subscription, then additional PT costs will be negligible. If, however, the chemistry laboratory must enroll in a new series of PT, the additional cost is a fixed indirect cost of testing.

There are several commercial sources for high quality proficiency testing samples, so it is wise to shop around if cost is an issue and the laboratory does not already receive a series which includes the analyte(s) in question. Your vendor should be able to supply you with all available sources of proficiency testing for the particular analyzer in question.

Before selecting a proficiency testing source, check the volume of subscribers that they have; it is always to the laboratory's advantage to enroll in a well-subscribed program, since statistical comparison improves as the number of subscribers increases. A list of approved Proficiency Testing Programs approved by HCFA is given in the index at the back of this chapter. If your vendor suggests a PT program that is not on this list, check with your area office of HCFA to see if it has been added.

LABOR COSTS

The previous section described in depth most of the material costs associated with performing a chemistry test. In this section, we will explore what is usually the most expensive direct cost component of a laboratory test--the labor component.

There are a number of ways to determine the labor cost of a particular chemistry test. However, before making such a determination, one should ensure that he or she uses a *fully-loaded hourly rate*. A fully-loaded hourly rate is the hourly salary for the testing personnel, plus benefits and productivity. Your Human Resources Department will be of help to you in calculating these numbers. To calculate a fully-loaded hourly rate, you must have the following cost components:

Salary

One of the most easily identified components of labor is the annual wage of the workers associated with testing. A study[6] conducted in 1992 revealed that medical technologists' starting salaries averaged $24,700 per year. To reduce the high cost of labor, many laboratories use personnel who have less technical training to perform routine tasks related to testing set up and other tasks. When determining a labor cost per test, do not forget to include all personnel actually involved with testing.

Benefits

Benefits offered by institutions vary widely and are generally considered to be a significant portion of the labor cost. Benefits commonly represent an increase over the base wage of 30-40%[7], though the rate can vary greatly from institution to institution.

The table below lists the most common benefits offered by employers.

<u>Categories of Employee Benefits (Source: Chamber of Commerce of the U.S., Employee Benefits 1983)</u>

I. Legally required payments (employer's share only):
   - A. Old-Age, Survivors, Disability, and Health Insurance (FICA);
   - B. Unemployment Compensation;
   - C. Worker's Compensation (including estimated cost of self-insured).

II. Pension, insurance, and other agreed-upon payments:
   - A. Pension plan premiums and pension payments not covered by insurance type plan (net);
   - B. Life insurance premiums; death benefits; hospital, surgical, medical, and major medical insurance premiums, etc. (net);
   - C. Salary continuation or long-term disability;
   - D. Dental insurance premiums;
   - E. Discounts on goods and services purchased from the company by employees;
   - F. Employee meals furnished by the company;
   - G. Miscellaneous payments (compensation payments in excess of legal requirements, separation or termination pay allowances, moving expenses, etc.).

III. Paid rest periods, lunch periods, travel time

IV. Payments for time not worked:
   - A. Paid vacations and payments in lieu of vacation;
   - B. Payments for holidays not worked;
   - C. Paid sick leave;
   - D. Payments for State or National Guard duty; jury, witness, and voting pay allowances, payments for time lost due to death in family or other personal

reasons, etc;

V. Other items:
   A. Profit-sharing payments;
   B. Contributions to employee thrift plans;
   C. Christmas or other special bonuses, service awards, etc;
   D. Employee education expenditures;
   E. Special wage payments ordered by courts, payments to union stewards, etc;
   F. Purchase of protective clothing (as required by OSHA);
   G. Payment for maintaining required employee license or certification.

Productivity

Most Human Resources departments recognize that an employee is not 100% productive 100% of the time; non-productive time must be considered, as well. Personal time for breaks and personal necessities, fatigue, and delays reduce an employee's productivity.[8] Your Human Resources Department may be helpful to you in determining a productivity factor, if your institution uses one, which will help you arrive at a fully-loaded figure to use in cost per test analyses. Productivity of 80-85% is commonly used.

Training

Another labor cost, as noted in the table, is the cost of training technologists. There are two main types of training costs: continuing education units (CEU's), and training programs for new instruments and/or methodologies. Another cost which might be included in this category is tuition reimbursement for college or other advanced degree coursework.

Full Time Equivalent (FTE)

The FTE is a management tool used in various labor calculations and is simply the number of budgeted hours for an employee (or group of employees) divided by 40, the number of hours in a work week. Thus two technologists who each work half-time comprise 1 FTE:

(20 hours/week from tech A / 40) + (20 hours /week from tech B / 40) = 1.0 FTE

If one wants to calculate the fully-loaded hourly rate for an employee, or for a group of employees, then, the formula is:

hourly salary x productivity factor x (1 + benefit factor) x FTE

If one wants to accurately determine the labor component of the cost of the test, one must determine a productivity standard for the test itself. In the laboratory world, the most commonly known productivity standards were originally set forth in 1970 by the College of American Pathologists in their widely used Workload Recording Methods.

The standards that the Method recommended were called *workload recording units,* which represented the minutes it took to perform a test as well as other activities that went into achieving high-quality results for that particular test. These included running controls, standards, and blanks, and performing routine maintenance on the instrument.

For a number of reasons, in 1993 CAP stopped supporting the Workload Recording Method, but many institutions, especially those that use *flexible budgeting* still have a need for a productivity unit per test.

Flexible budgeting occurs when the productive portion of the labor pool varies with the actual volume of work. Thus, in an environment where flexible budgeting is in place, technologists are not scheduled to work or are sent home if the work volume decreases. The accounting systems which track the budget are usually built "from the ground up", so a work unit per test is required. Activities which should be considered when determining a work unit per test[9] are listed below:

Components of Productive Time

I. Initial Handling of the Specimen
   A. Time stamping the requisition
   B. Sorting requisitions and specimens
   C. Recording the patient's name
   D. Assigning a laboratory accession number
   E. Entering information on a worksheet or computer terminal
   F. Labeling the sample
   G. Placing the sample into and removing it from a centrifuge
   H. Removing the serum/plasma
   I. Delivering the sample to the work area

II. Specimen Testing
   A. Diluting the specimen
   B. Adding reagents
   C. Adjusting the measuring instrument
   D. Placing the sample in the instrument
   E. Taking and recording readings
   F. Removing samples from the instrument

III. Recording and Reporting
   A. Calculating the result(s)
   B. Recording the result(s) on the patient's report and in the laboratory's records or in the computer
   C. Checking, sorting, and filing the final reports
   D. Taking incoming calls and making outgoing calls related to the initial report

IV. Daily and Periodic Activities
   A. Preparing quality control from lyophilized specimens
   B. Preparing bulk reagents
   C. Washing, drying, and sterilizing glassware
   D. Instrument cleaning and warmup
   E. Daily electronic calibration
   F. Instrument cleaning prior to shutdown

V. Maintenance and Repair
   A. Includes all standard maintenance procedures performed by laboratory staff at set intervals
   B. Includes minor repairs, part of which is defined as time spent identifying the defect

VI. Direct Technical Supervision
   A. Performing any indicated delta checks
   B. Reviewing quality control statistics
   C. Verifying and approving the reporting of results
   D. Supervising the procedure (not necessarily performed by a formally designated supervisor)

Note that time set aside for running controls, standards, repeats and blanks is not included in this list; those procedures should be counted like any other test as productive time.

OVERHEAD COSTS-SOURCES OF DIRECT OVERHEAD EXPENSE

The third major component of cost is overhead. Overhead includes all expenses which are incurred in the course of doing business, but which cannot be considered a direct cost of running a chemistry test. In a not-for-profit organization, the *contribution* which remains after variable expenses have been subtracted from revenues goes toward reducing overhead expense, and is usually referred to as contribution margin or contribution to overhead. The chemistry laboratory must contribute toward overhead expenses, just as all other departments of the laboratory and the institution do.

Overhead expenses exist in the infrastructure of the laboratory (referred to as *direct overhead expense)* as well as in the institution (*indirect overhead expense).* Frequently, all direct overhead expenses are not included in chemistry department's operating statement, and a chemistry manager must remember to include them when costing. Sources of direct overhead expense are described in this section.

Central Processing/Accessioning

If your facility has a central processing and accessioning department (CPA), it is usually an overhead area which is supported by the revenue-generating departments of the laboratory. While some revenues may be booked in a CPA department (such as those from phlebotomy or specimen processing), they are frequently insufficient for covering expenses incurred in the department. Although laboratories with a high degree of automation have been able to reduce their expenses in CPA, many laboratories still find that CPA is a labor-intense (and therefore expensive) department of the lab.

Reference Laboratory Expense

Though sometimes considered a "pass-through", where the expense is recovered in revenues, fees paid to a reference laboratory are nonetheless booked as an expense in your Accounting Department. Further, whereas just a few years ago, a laboratory could expect nearly full reimbursement for their reference laboratory expense, contracts and capitated plans now ensure that the laboratory recovers less than what they charge the patient or the insurer (such as when a plan pays for 60% of a billed lab test) and what they are charged by the reference laboratory. This means that reference lab expenses can be quite high.

In one recent study,[10] the researchers found that sendout expenses were the second-highest of all of their variable direct costs. As a result, many laboratories have requested that their reference laboratories share the risk and accept the payment that the billing institution receives. In certain cases, reference laboratories are billing patients or insurance companies directly. Before implementing a plan such as this to improve your recoveries, make certain that you are clear on laws regarding billing. Your own billing department should be able to help you.

Laboratory Information Systems (LIS)

The LIS section is also a source of direct overhead expense, depending on your system's size and complexity. An "off-the-shelf" system with no customization will be less costly than a complex system with a high level of customization. Remember when considering the LIS section's costs, to include labor, maintenance, and any off-site storage of tapes or drives for which your laboratory may pay.

Student Training Programs

Laboratories that participate in training students of medical technology or clinical pathology incur additional expenses. In larger programs, several FTE's may be devoted to teaching and training. Before agreeing to participation in a teaching program, a chemistry manager should have a clear understanding of its cost components. Development of teaching plans, preparation of learning materials, establishing a student "wet lab", evaluations of students and their work, and maintaining records and materials are all time consuming endeavors that contribute to the cost of a training program. The manager should carefully negotiate with the academic institution on the reimbursement the chemistry department might expect.

Outreach/Marketing

If your laboratory is involved in outreach or marketing, the activities that support these functions are direct overhead expenses. Customer service representatives, billing and collection clerks, contracted services to support the outreach program (such as an accounting service), couriers, vehicles, insurance, sales and marketing representatives and marketing expenses are all a part of this overhead. If you have a large marketing program, the direction and management which supports it is included as an overhead expense, as well.

Administration

Even very small laboratories usually have some administrative expense. Laboratory management is almost always included in this overhead category, and sometimes laboratory direction is included (again, check with your accounting department to determine how medical direction expense is handled in your organization). Receptionists, transcriptionists, and other clerical support personnel are included here; fees associated with licensure, accreditation, and certification are included here, as well. Some laboratories choose to include costs associated with compliance issues under this category - the rationale being that they are laboratory-wide expenses.

Other Laboratory Expenses

Depending on your own organization, there may be additional overhead expenses. Your laboratory may employ chemists who make reagents, or laboratory aides who maintain glassware. Your laboratory may have personnel dedicated to materials management, purchasing, and inventory. Very large, busy laboratories might employ compliance officers to help the staff understand and comply with regulatory issues. Irrespective of where a laboratory employee works, or what role he or she plays, all employees on the laboratory payroll must be included as a direct overhead expense when costing.

Just as Chemistry must cover its share of direct overhead expenses, it must also cover its share of indirect overhead expenses. Indirect overhead expenses can be thought of as expenses that the institution must spread across many departments and functions as a cost of doing business. Some of these expenses, such as disposal of hazardous waste, are obviously germane to the chemistry laboratory's activities, while others, such as support of a Community Relations Department, may seem less so. In most cases, however, the revenue-generating departments are expected to support all of the activities of an institution so that its goals and objectives are met. Some of the indirect overhead expenses are listed in this section.

Facility Overhead

If the clinical chemistry laboratory is part of a larger facility, there will be facility costs assessed to the department. These include such expenses as rent, (usually determined by the number of square feet that a laboratory occupies), water, electricity, gas, telephone, and insurance. The facility overhead may also include chemistry's share of the expense for entire departments such as Maintenance, Housekeeping, and Security.

Disposal of Hazardous Waste

Although this is usually a task handled through a Facilities Department such as Housekeeping, it's worth mentioning as a separate expense category because it can be so significant for a chemistry laboratory. Your department's income statement should provide a breakdown of what your department pays to dispose of its hazardous wastes; if it does not, ask your institution's financial manager for the information. Clinical chemistry laboratories frequently have at least three types of hazardous waste:

1. Biohazardous waste--blood and body fluid products and disposables that have come into contact with them;
2. Radioactive waste--usually fairly low level in a chemistry laboratory, but may be significant if your laboratory is involved with research;
3. Chemical waste.

If a chemistry manager or director does not know how his or her facility disposes of their wastes, it is important to find out. Under certain laws a director can be held responsible for improper disposal of waste; a large, costly fine is not usually budgeted and should be avoided! The director should seek out the Facilities Director or other responsible authority to learn about their own state or local laws.

Costs of Regulatory Compliance

We have included this category as an indirect overhead expense, but an argument could be made that it is a direct overhead expense for a chemistry laboratory. Use your best judgment about how to logically classify this cost in your institution.

The costs associated with regulatory compliance have skyrocketed in the last ten years. OSHA now requires that employers provide personal protective equipment, including non-permeable laboratory coats, to all employees working with blood or body fluids; most laboratories have interpreted

OSHA's *Bloodborne Pathogen Standard* as requiring gloves for every phlebotomy. The expense for vinyl gloves alone can run into tens of thousands of dollars for a busy laboratory.

If your laboratory has a dedicated safety or compliance officer, his or her time is an overhead expense. If, as in many laboratories, compliance is managed by bench techs or supervisors as part of their job, the time that they spend on safety and compliance issues is overhead time.

Writing, reviewing, and updating safety manuals and a Chemical Hygiene plan (required by OSHA), maintaining Material Safety Data Sheets (MSDS), keeping a chemical inventory, and maintaining records on chemical waste disposal are all time-consuming activities that can be a part of managing a chemistry laboratory.

Assigning an overhead category to costs associated with regulatory compliance may at times be confusing. In some instances, some or all of these tasks are managed by the overall facility; to cost out these activities, you must make a determination regarding under what department they fall. If they are handled within the laboratory, they are direct overhead costs; if they are managed outside of the laboratory, the are indirect overhead costs.

There are regulatory costs, however, that are more straightforward. Fees for mandatory licensure and accreditation, such as those for CLIA and state, are clearly direct overhead costs. Fees associated with voluntary accreditation, such as through the College of American Pathologists, are also direct overhead costs. However, if your organization pays an accreditation fee for the entire facility, the cost may be allocated across all departments and will appear in the indirect overhead line in your operating statement. When such a cost is allocated to all departments, it is usually negligible for the chemistry laboratory.

Decentralized Testing as an Overhead Cost

In the 1980's, decentralized testing became a major force in most healthcare institutions, but the passage of CLIA '88 left many decentralized testing sites in a turmoil. Decentralized testing, near-patient testing, point-of-care testing, and bedside testing are terms that are used interchangeably when talking about testing that is performed outside of a central laboratory area. Decentralized testing as used here can actually have two definitions:

1. The first definition applies to laboratory management of a satellite laboratory that duplicates services found in the central laboratory for the benefit of improved turnaround time in a critical nursing area, such as the Emergency Department.

2. The second definition of decentralized testing applies to point-of-care testing usually done by non-laboratory personnel, though some hospitals have staffed this function with medical technologists.

While many physicians believe that the availability of decentralized tests provides more immediate turnaround time, tests performed in this manner are frequently very high in materials cost (as well as labor cost). Further, the risk of operator error may be higher if the instruments or devices are used by someone who has not had proper training.

Many institutions have targeted decentralized testing as an unnecessary expense, and laboratories have responded by developing ways to improve turnaround time. The following section describes in detail types of decentralized testing and the authors' experience with them. We also provide a case study of how our laboratory responded to decentralized testing in its facility.

Bedside Testing: Point-of-care testing includes bedside testing. In some institutions, bedside testing is performed and managed by the laboratory; in others, as was initially the case in our own institution, the laboratory has no involvement with bedside testing. However, as we began the task of complying with CLIA '88, for the first time, members of Hospital Administration, the Department of Nursing, and the Clinical Laboratory met to review programs for managing bedside testing in such a way that patient care was not compromised and that the results obtained at the bedside were reliable. A Decentralized Testing Committee with members from each department was established to bring the various testing sites into compliance with CLIA '88.

The group began by administering a survey of what testing was being done on the various hospital units. The survey showed that by far, the most common bedside testing being done was blood glucose monitoring. Nurses were using whole blood glucose meters that had been approved by the FDA as suitable for sale to individuals for self-monitoring. Of course, these hand-held devices met the requirements for waived testing under CLIA '88.

The number of instruments in use was surprising; there were 105 glucose meters being used on various hospital units. On some of the more critical care units there were up to five instruments; a glucose meter could be left in the rooms of patients that were experiencing problems in glucose stability. Nursing had done a very commendable job in writing methods, training, and especially in recertification of competency, which was being done on a quarterly basis; however, no revenue was being recovered for any of the testing being done--nurses saw their roles as advocates for the patients and that included saving the patients money by not sending the tests to the laboratory where they would be charged.

The survey also revealed that tests were being performed at the bedside that did not fall into the waived category, such as blood gases in the operating rooms. For moderately and highly complex testing, there are very significant economic considerations in following CLIA '88 regulations. Proficiency Testing must be done. Even if the bedside site is considered part of the central laboratory and the instrument in the laboratory is considered the primary instrument for Proficiency Testing purposes, regular correlation studies with the primary instrument and other instruments in use must be done. Calibration or calibration verification must be done with each shipment of reagents even if they are of the same lot number previously used. An extensive quality control program similar to the chemistry laboratory's must be established, followed, and monitored--activities with which nursing units are not well prepared to deal. All of these requirements improve the quality of the test result, but create duplication of services with the chemistry laboratory and thereby increase costs.

It has been our experience that for moderate and highly complex tests done outside of the laboratory, meeting the CLIA '88 requirements is difficult and takes nurses away from primary patient care. As a result, most bedside testing sites in our institution have recognized that in many (but certainly not all) cases, the best place to do laboratory testing is in the laboratory. When quality patient care necessitates bedside testing, the nursing units work closely with the laboratory and do their best to meet the legal requirements.

While the Decentralized Testing Committee is certainly not finished with its work, there has been a spirit of teamwork between the laboratory and nursing, and much progress has been made. For example, in order to meet the turnaround times needed by the operating room (O.R.), a pneumatic tube is being constructed directly between the O.R. and the section of the laboratory where O.R. testing is being done. A computer workstation which sends test results to fax machines located in each of the O.R. rooms is used to communicate results immediately, temporarily bypassing the Laboratory information System. In the long run, this will cost much less than trying to maintain a laboratory in the O.R. As a result of this project, all O.R. blood gases have been moved to the laboratory.

Specialty Laboratories: Academic hospitals frequently have another source of competition from the viewpoint of the central laboratory--small "specialty" laboratories that tend to proliferate in a teaching hospital involved in research. The birth of a new laboratory usually goes something like this: a physician or department gets a research grant to explore some cutting edge technology. As part of the investigation, testing is done with informed patient consent but at no charge. For those tests that turn out to be beneficial in terms of diagnosis, monitoring the severity of a condition or efficacy of therapy, demand is generated by the physicians whose patients have been involved in the study. The tests are usually done at no charge as long as the research grant helps with the cost of getting the tests done and before the tests have received FDA approval; when the grant runs out, then the testing must be discontinued or a charge for testing must be generated (FDA approval must be obtained, as well). Usually the latter occurs, and a new laboratory is created.

Another option does exist: transferring the new test to the central laboratory. Often, this does not happen for a number of reasons. The physician who originated the test feels a significant amount of ownership for the establishment of the test, and may be unwilling to give up the newly established laboratory and test. Also, such esoteric testing is frequently not cost effective for the central laboratory--it is always expensive to bring a new test on board, the volume may be relatively low, the test may be too labor intensive, etc. Furthermore, the physicians who have done the research are generally less interested in the test's cost effectiveness than in its possible benefits to patient care, and are more interested in maintaining consistency in their research protocol. Thus, it becomes easier for an institution to perpetuate a new laboratory rather than reduce the costs associated with it.

Reference Laboratories: To a degree, reference laboratories can be considered a source of decentralized testing. Generally, they will perform tests at a much lower price than that charged by hospital laboratories. This is true for a number of reasons; their volume allows them to produce tests at a lower cost than most hospital laboratories; also, many hospital laboratories are left with comparatively high prices from bygone days when the laboratory was considered a revenue generator and was expected to support a significant proportion of hospital costs.

From a budgeting standpoint, it is difficult to lower pricing to compete with a reference laboratory. Administrators rarely want to hear about ways to reduce revenues through price-cutting, to compete in a market that they view as unimportant! If a hospital laboratory *can* re-evaluate their prices and set them at a competitive market rate it may be possible for hospital labs to serve as reference laboratories for nearby physicians who are currently sending their testing to more remote labs.

Despite this potential revenue boon, hospital labs must look hard at low volume and labor-intensive tests and consider whether it is really cost-effective to do those tests in house. Of course, the degree of medical urgency required for the results must be taken into consideration, but often turnaround time from a reference laboratory that performs tests every day may be shorter than that of in house testing that is done only a few times a week. Also it is important to consider factors such as expiration dates of reagents for infrequently run tests; do the reagents expire before they are used? How often must recalibration or patient correlations be done when the reagent lots (or shipments) change?

Usually for infrequently run tests, a minimal number of technologists are trained to do each test, but what happens when all of those techs are not available and a test comes in? If getting the test done requires a call-back of a technologist not working, it's an expense that must be considered.

Another factor to consider is the way to maintain a technologist's proficiency for a test that he or she performs infrequently. Repeat testing when the test run doesn't work adds to the expense of maintaining the test. If a tech has not run the test in a very long time, even if the test works it may take six hours rather than the usual four hours for the test to be run.

Point-of-Care Testing: Hospitals, laboratories, and suppliers to the healthcare industry have scrambled to respond to the rapidly changing forces described at the beginning of this chapter. One response that has greatly affected laboratory medicine is the proliferation of instruments designed for point-of-care testing. Manufacturers have aggressively marketed these instruments not only to laboratories but to every conceivable hospital unit as well; the instruments have opened an entire new market.

Companies have met with success in this marketing venture at least partly because one of the universal complaints about laboratory service has always been (and still is) turnaround time. In a stressful critical care situation, results from the laboratory are never available as quickly as the physician would like.

There has also been pressure on hospital laboratories to open ancillary laboratories in Critical Care units, such as the Emergency Department, Operating Room, and other intensive care areas. The reasons why such laboratories have appeal to such units is not difficult to understand; the close proximity of the lab allows direct communication between the physician or nurse and the technologist doing the tests. They can check to see when results will be ready and can see that the technologist is busy working to get the results. If the tech is working on another patient's stat orders, the physician knows why the result is not done, and can take action to negotiate with the other patient's physician to determine test priority. An *esprit de corps* is generated between laboratory staff and the unit staff.

In larger hospitals or clinics there are also political pressures which have a bearing on the appeal of such laboratories. Certainly the unit that has its own laboratory seems more important than other units which do not.

Following is a case study done from our own institution. You will see how total quality management was (and continues to be) effectively used to solve a problem concerning decentralized testing for the Emergency Department. As a result of the project described here, costs were greatly reduced and remain so, more than two years later.

Case Study--Point-of-Care Testing: A number of years ago, management decided that the hospital needed to open an adjunct to the Emergency Department (E.D.) that would compete with "Doc in the Box" emergency care clinics that had so successfully proliferated in nearby communities. The reasons for their success were their convenience and accessibility, shorter waiting time, lower prices, and often the "TLC" given to their patients: there was usually a follow-up call the next day from the Physician or Physician's Assistant to the patient to check on their progress.

In preparing our own Emergency Department for prompt care, one of the steps taken was that of establishing a Prompt Care laboratory within the E.D. (it was believed that this lab was necessary to compete with the new local clinics, even though most of them had only a very basic laboratory). Most laboratory results were called to the patient the next day as part of the follow-up procedure.

The space allocated for the laboratory was very small and somewhat dictated the choice of instrumentation used in the lab. The chemistry instrument chosen, while quite compact, was much slower than the fully automated instruments used in the main lab. The lab was to be staffed by one person, and since at that time all of the technologists were specialized in either chemistry or hematology, a major retraining effort was required to provide staffing for the new laboratory.

The number of Prompt Care patients was low, so the Prompt Care laboratory very soon became the Emergency Department laboratory in everything but name. From the beginning, the laboratory was a major problem for laboratory supervisors, who were trying to schedule their staff efficiently. Since it was located quite some distance from the main laboratory, coverage for meals and breaks was a problem in trying to anticipate a time when there would be an easy turn-over of duties from one technologist to another. Very often the tech to be replaced was in the middle of doing a manual differential and the replacing technologist had to wait until it was finished.

There were many lag periods when no samples were received and after daily maintenance was completed, the tech had nothing to do. It became increasingly more difficult for supervisors to think of useful projects to keep the tech occupied. The limited space precluded working on development of new tests or instruments; it was not unusual for the tech to be sitting and waiting for a specimen. Of course, techs were encouraged to read technical journals and work-related material during such times but supervision by long distance is difficult and many crossword puzzles were done.

At other times, more than one specimen arrived in the lab at once and the technologist was overwhelmed. The chemistry instrument was slow and the tests for the electrolyte and renal panels had to be done one at a time. Tests were often ordered which could not be done by the lab's limited instrumentation; in that case the specimen had to be processed and part of the sample sent to the main laboratory. The Emergency Department usually ordered slide differentials and there was no way for the technologist to speed up that process.

When very unusual cells were seen or when the technologist was one whose primary experience had been in chemistry, it was often necessary to send the slides to the main lab for review by another, more experienced technologist. Conversely, when there were problems with the chemistry analyzer and the tech assigned was one whose primary experience had been in hematology, it was often necessary for a supervisor to try to help with troubleshooting by phone or to send a more experienced person to the Prompt Care laboratory.

As a result of the alternating boredom and being overwhelmed by more specimens than one could handle, the assignment to be the prompt care technologist was a very unpopular one and did not help morale in the laboratory.

In December of 1991, there were complaints from the Emergency Department management about turnaround time from their Prompt Care Laboratory. The Q.A. Coordinator from the laboratory did a preliminary turnaround time study, which confirmed what was already obvious to the laboratory: as long as specimens arrived in the Prompt Care lab evenly spaced in time, the turnaround times were very good but when they arrived in batches, the times were not acceptable.

The Emergency Department loved having their own lab and the convenience of talking to the technologist directly; the solution they proposed was to enlarge their laboratory and buy faster and more automated instruments comparable to those in the main laboratory. The costs involved would have been enormous since there was no space available and there would have to have been major remodeling and moving costs to free up sufficient space for the expanded laboratory, not to mention the cost of the new instruments.

A Total Quality Management (TQM) Committee was formed to further assess the problem. The members consisted of the Laboratory Manager, the Laboratory Chief of Operations, the Laboratory Q.A. Coordinator, the Nurse Manager from the Emergency Department and the Chief Resident of the E.D. This group met weekly to discuss possible solutions to the problem. The laboratory was anxious to find a way to meet the turnaround times required; the committee members from the laboratory believed that improvements in turnaround time could be achieved by changing the way they handled specimens in the main lab.

The Emergency Department agreed to give it a try for a three-month pilot, as long as there were weekly meetings of the committee and that the laboratory would agree to terminate the experiment immediately if there was a deterioration of turnaround times.

A turnaround time goal of a thirty minutes was established and the two departments negotiated and agreed upon which tests were truly needed in that time frame. A one month statistical study of pre-pilot turnaround times was done as a baseline for the tests involved, and the pilot began.

The laboratory used the FTE previously assigned to the Prompt Care laboratory to fill a newly created position called E.D. Coordinator. That person's responsibility was to receive specimens from the E.D., accession the specimens into the laboratory computer, do whatever processing was necessary (such as centrifuging chemistry specimens), deliver the specimens to the appropriate departments, and ensure that the technologists in those sections knew the specimens were from the E.D. and were given the highest priority. It was also the E.D. Coordinator's responsibility to check on the progress of those tests in the computer and to ensure that there were no delays.

The Laboratory Q.A. Coordinator did a daily study of turnaround times and those statistics were reviewed by the Chief of Laboratory Operations, the laboratory supervisors of all participating sections, and the E.D. Coordinator. Any problems that occurred were addressed on a daily basis and at weekly Supervisors' meetings. However, the number of problems that occurred were few.

When there was more than one manual test to be done such as differentials, monospots or rapid strep tests, they could be given to different technologists. The chemistry tests could be done more quickly by the automated instruments in the main laboratory. Other tests which had not been available in the Prompt Care Laboratory such as coagulation tests and an expanded menu of chemistry tests could also be done much more efficiently.

Weekly statistics of turnaround times were plotted against the data in the pre-pilot study. The data and the graphs were shared at the weekly TQM meetings. At each meeting the E.D personnel expressed their amazement at the improvement in laboratory service. At the end of three months, all of the data was collected and plotted against the pre-pilot data. The data for the post-pilot period was collected for three months and the data for the pre-pilot period was collected for one month.

The chemistry data from the E. D. Pilot March-May 1992 is given in Fig. 1. Because the pilot was such a success, the Prompt Care laboratory was closed and the E.D. and the laboratory had a party together to celebrate their shared accomplishments.

Fig. 1. **CHEMISTRY DATA FROM E.D. PILOT**
MARCH - MAY 1992

ELECTROLYTE PANEL	RENAL PANEL	TOTAL CK	MAGNESIUM	AMYLASE
MINUTES	MINUTES	MINUTES	MINUTES	MINUTES
ALL SHIFTS AVERAGE = 19.50	ALL SHIFTS AVERAGE = 19.45	ALL SHIFTS AVERAGE = 25.39	ALL SHIFTS AVERAGE = 23.96	ALL SHIFTS AVERAGE = 24.64
NUMBER = 1557	NUMBER = 1156	NUMBER = 254	NUMBER = 121	NUMBER = 248
MIDNIGHT SHIFT AVERAGE = 23.18	MIDNIGHT SHIFT AVERAGE = 23.10	MIDNIGHT SHIFT AVERAGE = 26.98	MIDNIGHT SHIFT AVERAGE = 24.56	MIDNIGHT SHIFT AVERAGE = 25.67
NUMBER = 283	NUMBER = 210	NUMBER = 61	NUMBER = 32	NUMBER = 58
DAY SHIFT AVERAGE= 24.07	DAY SHIFT AVERAGE= 23.65	DAY SHIFT AVERAGE= 27.01	DAY SHIFT AVERAGE= 24.94	DAY SHIFT AVERAGE= 28.76
NUMBER = 535	NUMBER = 390	NUMBER = 94	NUMBER = 32	NUMBER = 82
EVENING SHIFT AVERAGE= 14.78	EVENING SHIFT AVERAGE= 15.12	EVENING SHIFT AVERAGE= 22.86	EVENING SHIFT AVERAGE= 21.95	EVENING SHIFT AVERAGE= 20.95
NUMBER = 739	NUMBER = 556	NUMBER = 99	NUMBER = 57	NUMBER = 108

The Turnaround Time Plot for hematology and chemistry done in the Prepilot study (Jan. 1992) Pilot study (March - May 1992) and Post Pilot Study (1993) is shown in Fig.2. There were no cost savings in terms of *eliminating* personnel since the FTE that had been the Prompt Care Tech became the E.D. Coordinator in the laboratory; however that person was now located within the lab and could help in many areas when no E.D. specimens were submitted. During the beginning of the pilot, a technologist was assigned to that position. As soon as the logistics for doing the task most efficiently had been established, the duties were transferred to a laboratory assistant in the specimen processing area so there *was* a cost savings for personnel.

As with any TQM project, monitoring the progress of the newly established process continues--the turnaround times have been kept on a monthly basis during the post-pilot period. The graphs are shown in figure 3.

We wanted to share our experience about improved turnaround times because closing a point-of-care ancillary laboratory seems to be counter to what is happening in many hospitals, and yet the factors that made our Prompt Care Lab inefficient and costly would seem to be universal for any point-of-care laboratory.

Costing Example for Clinical Chemistry- Cost/Test

Determining Costs Per Test: Now that we have defined the components which comprise total cost, we can look at ways to determine costs per test.

At one time it was common practice to attempt determining costs "from the ground up", that is, by trying to identify the cost of each of the material, labor, and overhead components that were discussed in the last section. Thus, a supervisor or manager would spend a great deal of time trying to calculate at the test level what the cost of the instrument, reagents, disposables, controls, standards, proficiency test material, and all of the various overhead expenses were.

Fig. 2. **TURNAROUND TIME PILOT EMERGENCY DEPARTMENT/LABORATORY**

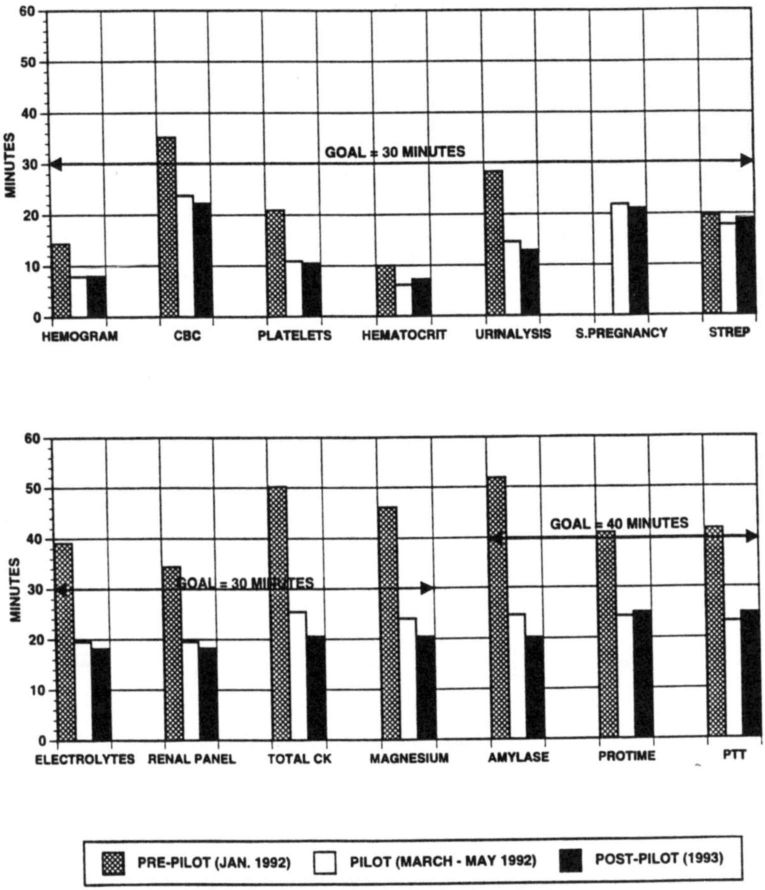

Fig. 3. EMERGENCY DEPARTMENT TURNAROUND TIMES 1993

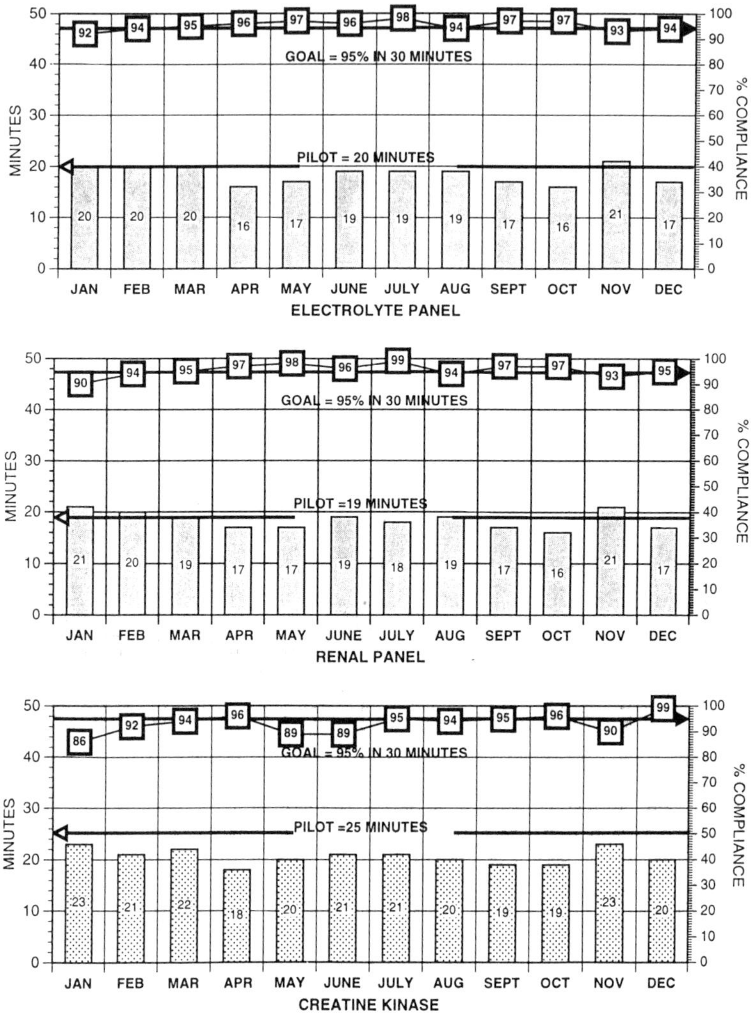

Of course, if the laboratory participated in CAP Workload Recording, the manager knew how much labor time went into each test, but would then have to figure out how much that time was worth in labor costs. Even if done very carefully, the final determination was a tedious process, and ultimately was inaccurate. for example, it was easy for a manager to forget to include the more peripheral, indirect costs such as disposal of biohazardous wastes.

At some point, assumptions had to be made about unusual situations like repeats, failed runs, down time, backup methods, and so on. Most managers and supervisors have come to recognize that the time and tedium involved in using this method to calculate cost per test are rarely more accurate than a weighted average method using different laboratory sections, different instruments (or workstations), and/or different test groupings. Usually by using a weighted average method, a manager can arrive at a reasonable cost per test for the vast majority of the tests available, especially in a highly automated area such as Chemistry. There may be a few tests that are exceptions (such as when a reagent kit costs significantly more than others used on a particular instrument).

Before beginning a cost analysis, the wise supervisor or manager will carefully assess the significance of a trade-off between a time-consuming, tedious method which puts a test cost at say, \$4.80, versus a quicker weighted average that determines the cost to be \$4.92.

The following example for costing a chemistry test helps illustrate how all of this information is used to determine cost per test. Follow the example along on the three worksheets.

The Laboratory Manager of City Hospital has been informed that the Finance Department will be installing a new accounting system. They have asked the manager to determine the cost per test for all tests, numbering more than 400 in all. They have given her four weeks to complete the task, but the manager's schedule is full and she is attending off-site strategy meetings for four of the available days. It is important for her to simplify the task without giving Finance grossly incorrect numbers. The manager decides to start with the Chemistry section for her analysis. She pulls out her most recent Operating Statement from the Finance Department and her most recent productivity data from the LIS.

On the Operating Statement (Worksheet 1), labor costs are divided between variable and fixed. Other expenses, however, are not as clearly delineated; it may be useful to determine which line items are fixed and which are variable. For now, however, our Lab Manager is going to look at "the big picture".

The Workload Records (Worksheet 2) include a sectional breakdown of unit volume and workload unit volume for the three areas within the Chemistry laboratory. On this report, City Hospital's LIS captures non-billable tests–very important since non-billable tests represent nearly 20% of the total testing volume. Cost per test calculations should always account for non-billable tests since they are a significant portion of the chemistry laboratory's volume.

The Laboratory Manager prefers to use the year-to-date (YTD) data since it includes nine months of data. Obviously, from a statistical perspective it is always better to include as many data points as possible in costing calculations. On Worksheet 3, she starts by performing the most simple cost per test calculation: total expenses divided by total number of tests. Again, when calculating this number be certain that you are using the total direct expenses divided by the sum of patient tests and non-billable tests. The average cost per test is $7.65. The Manager feels that $7.65 is a reasonable starting point, but she knows the tests in Special Chemistry and Toxicology are more labor intensive than those in the Routine Chemistry section, and she wants the cost per test data to reflect that difference.

WORKSHEET 1

City Hospital
Operating Statement--Chemistry Department
September 1994

	Sep-94 Actual	Sep-94 Budget	Sep-94 Variance	YTD Actual	YTD Budget	YTD Variance
Revenue						
Inpatient revenue	$2,073,609	$2,101,125	-1.3%	$17,936,718	$18,489,900	-3.0%
Outpatient revenue	$455,400	$385,874	18.0%	$4,007,520	$3,550,041	12.9%
Discounts allowed	($47,741)	($44,639)	6.9%	($439,217)	($397,287)	10.6%
Allowances	($36,552)	($35,495)	3.0%	($328,968)	($314,131)	4.7%
Contractuals	($1,625,517)	($1,606,518)	1.2%	($13,816,895)	($13,976,707)	-1.1%
Bad debt	($60,540)	($57,414)	5.4%	($520,644)	($499,502)	4.2%
TOTAL	$758,659	$742,933	2.1%	$6,838,514	$6,852,315	-0.2%
Expenses						
Fixed labor cost:						
Mgmt & Supervisors	$22,162	$24,631	10.0%	$195,026	$220,447	11.5%
Technical	$47,264	$55,987	15.6%	$415,923	$496,605	16.2%
Clerical & Support	$6,377	$7,884	19.1%	$55,480	$71,744	22.7%
Variable labor cost	$94,711	$78,878	-20.1%	$842,928	$721,734	-16.8%
Pathology fees	$45,520	$44,576	-2.1%	$414,228	$409,207	-1.2%
Medical gases	$1,820	$1,996	8.8%	$16,562	$16,770	1.2%
Radioactive materials	$234	$484	51.7%	$2,153	$2,305	6.6%
Other medical supplies	$445	$389	-14.4%	$4,005	$3,454	-15.9%
Minor medical equipment	$2,475	$1,215	-103.7%	$14,587	$10,619	-37.4%
Laboratory supplies	$98,745	$95,447	-3.5%	$879,654	$858,069	-2.5%
Xfer from Central Supply	$2,582	$3,699	30.2%	$28,238	$33,291	15.2%
Reference lab expense	$32,458	$30,145	-7.7%	$298,165	$304,514	2.1%
Depreciation	$3,897	$4,253	8.4%	$35,852	$38,107	5.9%
Office & admin supplies	$154	$185	16.8%	$1,386	$1,658	16.4%
Repairs & maintenance	$4,225	$3,997	-5.7%	$38,448	$35,973	-6.9%
Dues & subscriptions	$400	$200	-100.0%	$1,205	$1,400	13.9%
Travel	$0	$0		$0	$0	
Rent/lease equipment	$2,750	$2,750	0.0%	$24,888	$24,570	-1.3%
TOTAL	$366,219	$356,716	-2.7%	$3,268,727	$3,250,467	-0.6%
Contribution to Overhead:	$392,440	$386,217	1.6%	$3,569,788	$3,601,847	-0.9%

WORKSHEET 2

City Hospital
Workload Records--Chemistry
September 1994

	Sep-94 Unit Volume	Sep-94 WLR Units	YTD Unit Volume	YTD WLR Units
Total number of patient tests--Routine Chemistry	27,487	82,461	246,009	738,026
Total number of patient tests--Special Chemistry	10,931	76,517	97,067	679,471
Total number of patient tests--Toxicology	1,255	15,060	11,295	135,540
Non-billable tests (controls, standards, blanks, repeats)				
Routine Chemistry	5,552	16,657	49,749	10,049
Special Chemistry	2,285	15,992	20,333	4,250
Toxicology	314	3,765	2,855	714
TOTAL	47,824	210,452	427,308	1,568,050

The manager knows (and confirms by calculation on Worksheet 3) that labor is a significant portion of overall expense--46%. She feels that if she isolates variable labor costs in her calculations, the total costs per test may change significantly. By using a weighted average calculation with year-to-date Workload Records Units (WLR Units) (Worksheet 2), she calculates the percentage of the total workload units each chemistry section requires, and comes up with 47.7% for Routine Chemistry, 43.6% for Special Chemistry, and 8.7% for Toxicology. Using these percentages, she calculates weighted average variable labor cost per test, as demonstrated in our example on Worksheet 3. She then subtracts variable labor costs from the total costs, and uses the remaining number to calculate all other costs per test.

Now, instead of submitting to Finance one cost per test for all of chemistry, the Laboratory Manager can submit at least three costs per test, depending on the section in which the test is performed. If our Laboratory Manager has sufficient time, she can further refine her calculations by breaking out variable supply costs and other variable costs and performing similar calculations, or by stratifying the Chemistry sections into different workbenches and recalculating.

While this example is useful for demonstrating a "quick and dirty" method of deriving cost per test, one must use logical judgement in calculating all costs per test like this. If City Hospital, for example, does a large volume of routine chemistry and electrolyte panels on the same workhorse instrument, it would most likely be useful to determine how much the materials cost per test is for tests performed on the instrument. Your Accounting Department should be helpful in providing dollar volumes paid to a particular vendor; if all of the reagents, standards, and controls are purchased through the instrument manufacturer, then it is a fairly safe assumption that variable supply cost can be calculated from these data. Frequently however, these exact same types of exercises turn out to be the most useful when the Manager is under the pressure of a tight deadline.

WORKSHEET 3

Manager's Worksheet for Calculating Cost per Test:

Average Cost per Test Analysis:

Total YTD Cost (from Operating Statement) / Total YTD number of Tests (from Workload Records) =
$3,268,727 / 427,308 =
$7.650

Percentage of Expense Attributable to Labor:

Total Labor Expense (from Operating Statement) / Total Expense (from Operating Statement) =
$1,509,356 / $3,268,727 =
46%

Weighted Average Cost per Test Analysis:

Percent of Total Billable and Non-Billable WLU's per Section:

Routine Chemistry	(738,026 + 10,049) / 1,568,050 =	47.7%
Special Chemistry	(679,471 + 4,250) / 1,568,050 =	43.6%
Toxicology	(135,540 + 714) / 1,568,050 =	8.7%

Weighted Average YTD Variable Labor Cost per Test Analysis:

Routine Chemistry	.477 x $842,928 = $402,077	$402,077 / 295,758 = $1.356
Special Chemistry	.436 x $842,928 = $367,516	$367,516 / 117,400 = $3.130
Toxicology	.087 x $842,928 = $73,335	$73,335 / 14,150 = $5.180

Total YTD Costs - Variable Labor Cost =
$3,268,727 - $842,928 =
$2,425,799

Total YTD Cost - Variable Labor Cost / Total YTD number of Tests =
$2,425,799 / 427,308 =
$5.677

New Total Cost per Test, Accounting for Weighted Average Variable Labor Cost per Test:

Routine Chemistry	$1.356 + $5.677 = $7.033
Special Chemistry	$3.130 + $5.677 = $8.807
Toxicology	$5.180 + $5.677 = $10.857

Negotiation: When it comes to acquiring a new test system, it is always wise to represent your laboratory as being interested in a minimum of two comparable instruments when talking to sales representatives, even if there really is only one that you are seriously considering. There needs to be some semblance of competition to get the very best price that you can. Negotiating sales agreements is a skill that many technical personnel do not have; the novice may feel more comfortable recruiting a more seasoned negotiator for help. The technical supervisor must work carefully with the administrative negotiator because the negotiator must have a very thorough knowledge of all the technical issues involved.

As described earlier, the first step of a negotiation is clearly defining the needs of the clinical chemistry laboratory and all interested parties (e.g., the clinical chemist, the pathologist, the administrator). For a major equipment purchase this step will probably take a considerable investment in time; for example, last year our laboratory chose a new Laboratory Information System (LIS). The homework phase consisted of weekly meetings of the Laboratory Technical Advisory Committee comprised of one supervisor from each laboratory department, the LIS manager, a Ph.D. chemist appointed by the Laboratory Medical Director, the Quality Assurance Coordinator, and the Laboratory Administrative Manager.

Nearly a full year was spent in evaluating and comparing seven laboratory computer systems, eventually narrowing the choices to three systems. In-house demonstrations of the three systems were attended by a significant proportion of the laboratory staff. The Committee visited other hospitals that were using the systems. Many phone conversations with other hospitals using the systems were placed, and so on. All of this occurred before the negotiation phase began. The actual negotiation was done by the Laboratory Administrative Manager and the Associate Director of Hospital Information Systems (HIS), who was experienced in negotiating large information systems.

The Art of Negotiating by Gerald I. Nierenberg is

considered by many to be the classic text on negotiation. It is an excellent text and we have chosen some excerpts of interest here.

In Nierenberg's the chapter "Preparing for Negotiation", after discussing the need for doing your homework in determining your own needs, he states:

"It is essential to examine the opponent's past history, inquire into previous transactions he was connected with, and look into every business venture or deal he has consummated. Also investigate any deals he has failed to conclude successfully. Frequently you will learn as much, or more, about people from their failures as from their successes."[11] When doing our homework for acquisition of the LIS, we *did* gain valuable information from talking to our counterparts in other laboratories who had recently installed new laboratory information systems, by asking them why they had not chosen the systems that we were considering.

Interestingly, Nierenberg refers to the person with whom he is negotiating as the "opponent." The negotiating process is likened to at least a chess game if not an all-out war. In the chapter on "Negotiating Techniques" he discusses strategies and tactics. The following are listed as variations on the "When Strategy": " forbearance, surprise, fait accompli, bland withdrawal, apparent withdrawal, reversal, limits, and feinting."[12]

Many of the examples that he uses are appropriate for combat. The problem in using such tactics is that in situations such as those of negotiating with an LIS supplier, the association with the individuals involved will not be over after the negotiation process is finished; in fact, the association is just beginning and you want to set the stage for an atmosphere of cooperation and mutual success.

In such situations, the win/lose approach is more applicable by applying the "good guy/bad guy" tactic, where someone who is not vested in the project can apply the "winner take all" viewpoint. This was much the role that the

Associate Director of HIS played in our negotiation with the LIS vendor.

The Harvard Negotiation Project which began in the 1970s presented one of the first departures from the standard win/lose philosophy. In *Getting to Yes* by Roger Fisher, William Ury, and Bruce Patton, authors who were involved in the Harvard Project, a method called principled negotiation or negotiation on the merits, is advocated.[13]

There have been many books written on "win/win" negotiation techniques; many of these are just the standard "win/lose" approaches in disguise.

Added Value Negotiation,[14] by Karl and Steve Albrecht, is a refreshing text that advocates an approach in which the negotiation is conducted in a cordial and stress-free environment, characterized by openness, honesty, cooperation, empathy, and mutual trust. The focus looks at the interests of both parties and arrives at an agreement that is balanced and has satisfactory value for both. This approach reduces the adversarial environment by using a structured process to focus on the elements of value important to both parties. This includes working on multiple deal packages that include various combinations of the identified value elements, selecting the package that has the greatest appeal for both parties, and ensuring that the agreement selected has elements that go together into an effective solution for all concerned.

In the next section, we apply some of the principles of negotiating directly to the chemistry laboratory and equipment acquisition.

III. REDUCING MATERIAL COSTS THROUGH COST ANALYSIS

Earlier we gave an example of a cost analysis for a chemistry department. Especially in equipment acquisition, it is important to perform a thorough analysis to avoid pitfalls that may prove to be costly later. In this section, we will provide tools to help you make a cost-effective decision in acquiring a test system.

TOOLS FOR DECISION-MAKING IN CAPITAL ACQUISITION

When negotiating a capital acquisition with a vendor representative, remember always that his or her primary goal is *to sell you an instrument and/or reagents at the highest price that he or she can get.* They are usually commissioned salespeople; the more you pay for an instrument, the bigger their take-home pay becomes. They are frequently rewarded through bonuses for having a high number of placements (i.e., sales, leases, or reagent rentals), so they are primarily motivated to place as many high-priced instruments as they can.

Except in rare circumstances, do not mistake a sales representative's warm and friendly manner as anything but a sales tactic--they may want your friendship but they positively cherish your sale. It is not a coincidence that your sales representatives are usually attractive, friendly, bright people filled with exuberance--that is exactly the type of person that corporations seek for their sales positions. These are the people who can close the deal; who can achieve sales quotas. Sales quotas, by the way, are usually the reason a sales representative pressures a laboratory supervisor or manager to get a purchase order number within a specific short time frame, or why they may offer better deals when their fiscal quarter or year is about to end. It can be advantageous for a supervisor or manager to know what the vendor's fiscal calendar looks like, and most representatives will provide that information.

Additionally, though already stated, in making an

acquisition decision we cannot emphasize enough the importance of keeping competition alive as long as you can. Representatives become far less motivated to deal if they believe (or if they know with certainty) that they have the sale. Further, they must believe that price is a determining factor in the decision (as it well should be!).

Do not disclose how much you have to spend in your budget; no representative will be interested in helping you save money if he or she knows your budget--remember, they are paid on commission! If you are working with a team of people in your laboratory, agree before the acquisition process begins that only one person will negotiate the price, and that no one will "show his or her hand" to representatives from the company by telling them that their test system is the preferred system. Team members should also be instructed to avoid discussions regarding the instrument evaluation with their colleagues at neighboring institutions; otherwise the vendor may learn of your laboratory's acquisition intentions in this circuitous manner.

Let us be quick to point out that while it is important to keep competition alive, it is unprofessional to share price quotes from competing companies with vendor representatives. Though it does happen, the professional laboratorian should not be involved with providing to a vendor's competition cost proposals, price lists, invoices, or any other competitive information that a vendor representative may have given you. Do not let a sales representative use pressure tactics to obtain their competition's price quote (e.g., "How can I beat their price if I don't know what their price is?")--simply emphasize to him or her that you need their very best price.

With that in mind, it is important that the supervisor or manager obtain as much information about a capital acquisition as he or she can from the vendor. You should expect to provide the vendor with your monthly or annual unit volume for the tests in question, and tell the vendor representative that in return you want a cost analysis that includes the following elements:

CHECKLIST FOR VENDOR COST ANALYSIS

- Their list price for an instrument, reagents, and supplies;
- Their best price to you for the instrument--including analyses for outright purchase, lease, and reagent rental;
- Tax, shipping, insurance, and any other miscellaneous charges that you may have to incur (e.g., some companies want you to buy a "start-up kit" which contains parts and tools needed for your technologists to make minor adjustments and repairs);
- The annual cost to you over five years for reagents and disposables. Request that the recurring cost information be given in both cost per billable test and in units of sales, i.e., cases, boxes, each, etc;
- The frequency with which the manufacturer expects controls and standards to be run;
- The cost of controls and standards (when applicable);
- The five-year cost of an annual maintenance contract, and answers in writing to all of the following questions: What is included in annual maintenance? Will the contract cover replacement parts? Is an extended warranty (where the laboratory pays for say, three years of maintenance in advance) available with a discount? What hours will the maintenance contract cover? Is 24-hour customer service available? What is the company's guaranteed response time? Are parts available to independent contractors who provide biomedical engineering services? If your facility is large enough to support its own biomedical engineering department, it may be more cost effective for you to use them, if they can get the training and the parts.
- The cost of other capital equipment that may be required for the effective operation of their instrument--refrigerator, freezer, centrifuge, barcode printer, etc. Although your vendor may not actually supply these items, recruit their help in estimating your cost to purchase them.
- Facility requirements--dedicated electrical lines,

deionized water sources, removal of cabinetry, etc. If the new instrument has certain installation requirements, you should get a price quote from your Facilities Department or contractor on the cost of the changes required. As mentioned before, they can be quite expensive if major changes must be made. You should also arrange a meeting with your facilities engineer and the vendor representative, so any heating or ventilation requirements are clearly determined up front.

- Specifications for computer interface--either bi-directional or unidirectional, depending on what the instrument can do and what the chemistry laboratory's needs are. These specifications will need to go to your own programmers or to those of your computer vendor for a cost and time proposal.
- The frequency and typical cost for a software upgrade for the instrument. Will such an upgrade happen during the life of the instrument that you are considering purchasing? Can the company provide you this upgrade at no cost?
- Costs for training personnel--travel, hotel, training costs, etc.

This list is lengthy but all of this information is required to make an informed decision. Please be aware, too, that the cost of virtually every item on this list is *negotiable with the vendor*. Do not be afraid to ask the vendor to "throw something in" with the purchase price. Their cost for a new freezer may be far less than what you might pay, and their cost is usually low compared to what they will make on the sale of the instrument to you.

Once you have the vendor's price quotes, you are ready to do your own analysis. To complete a thorough financial analysis, follow these steps:

1. Review the data provided by the vendor. Did they use your volume numbers correctly, and are their calculations correct? You should be able to duplicate their work, and the steps that you must take to do so should seem logical. If you cannot reconcile their numbers, discuss the

analysis with your vendor representative.

2. Consider performing a breakeven analysis. A breakeven analysis will give you the number of tests that you must perform on the analyzer at the quoted costs to become profitable. To determine the breakeven point for a particular analyzer, use the following formula:

Breakeven Point = Fixed Costs / (Price per Test - Variable Costs)

In the case of analyzing the breakeven point for a particular instrument, derive the costs as follows:

Fixed costs--the sum of all costs that you would have to pay even if you did not run one test on the instrument. This would include such costs as depreciation, maintenance contract, and direct overhead costs.

Variable costs--add all costs that are associated with running tests, and that increase and decrease with volume. These costs include reagents, supplies, disposables, controls, standards, and labor (if your institution uses a flexible budget).

Example. You are considering buying an instrument that costs $150,000. At what volume of tests will the instrument turn a profit?

a. Calculate fixed costs:

Straight-Line Depreciation = Cost of Capital/Life of Instrument
= $150,000 / 5 years
= $30,000

Annual maintenance contract = $12,000

Direct overhead costs = 5% of total chemistry laboratory costs
= $50,000

Total fixed costs = $30,000 + $12,000 + $50,000 = $92,000

b. Variable costs (provided by vendor in cost analysis, reviewed and verified by you):

Variable cost per billable test = $6.00

c. Price per test = $16.00

Breakeven = $92,000/($16.00 - $6.00)
= 9,200 tests

Thus, performing the test on this instrument becomes profitable at any volume of tests greater than 9,200.

3. Perform a net present value analysis (NPV). From a strictly financial viewpoint, acquisition of capital equipment involves an investment of funds for the purpose of generating cash at a future time. The purpose of discounted cash flow analysis is to determine whether the acquisition is worth more to the institution than it costs.[15]

Besides net present value, other methods for assessing capital equipment acquisition are frequently used, most commonly payback and internal rate of return. While not without their merits, both methods have limitations in assessing an acquisition, and net present value analysis is generally viewed as being the best method of the three. It is not unusual, however, for financial managers to use two or all three methods for assessing an acquisition's worth.

In a net present value analysis, cash flows are determined over the life of the instrument. "Cash flows" mean all money that is spent on the instrument, as well as money (revenue) collected. Cash flows include all capital expenditures incurred at the initial phase of the acquisition (usually referred to as period 0), as well as gains from capital sales (such as cash from the sale of an old analyzer). Cash flows also include the revenue expected from the new instrument.

If the new instrument is replacing an existing instrument with no change in test menu, historic data is usually sufficient for determining future cash flows; if the new instrument allows the laboratory to bring up new tests, you will need to forecast incremental revenues.

For the final period of cash flow, the after-tax salvage value of the instrument should be included. Remember that cash flow determinations include only actual dollars that come into or go out of the laboratory or institution; do not make the mistake of including accounting allocations such as depreciation dollars or overhead. Use a spreadsheet package to set up your cash flows; you will want the spreadsheet to do the calculating for your net present value. The following example shows a typical cash flow analysis for choosing between two chemistry analyzers.

CASH FLOW ANALYSIS EXAMPLE

The laboratory manager, the chemistry supervisor, and the pathologist are deciding between two upgrades, Analyzer A and Analyzer B, for their immunoassay area. Analyzer A will perform all of the tests in question, but is too costly for the laboratory to purchase outright.

The sales rep from Company A has worked out a leasing arrangement where the lease cost is buried in the reagent and materials cost, meaning no upfront capital costs for the laboratory (except for writing the computer interface).

Another advantage to Analyzer A is that material pricing will be constant for the duration of the agreement, meaning no price increases in the coming years. However, Analyzer A is slightly more labor intense than Analyzer B, a major consideration for the purchasing team in these times of constant cutbacks.

Analyzer B, on the other hand, falls within the laboratory's capital budget, therefore allowing them to purchase the instrument. The problem with analyzer B is that the pathologist is not comfortable with the current methodology for some of the tests, and if analyzer B is the winner, he will want to continue performing those tests by the laboratory's current RIA method. The remaining tests, however, are less labor intensive than Analyzer A and material costs are significantly less.

The Laboratory Manager told the sales rep from Company A that the costs for Company B's analyzer were less; the Company A rep responded by offering the laboratory 6 months of reagent for $1.00, and reducing the annual maintenance contract cost to $10,000. The chemistry supervisor and the pathologist are now looking to the lab manager to make a cost-based recommendation. The different variables make it a complicated decision.

Table 1 estimates the annual volume of immunoassay tests in question, numbers that will be used in subsequent calculations.

Table 1.
Reagent and QC Costs per Billable Test for Community Hospital Immunoassay Testing Upgrade

Annual Volume:					
Test	No. tests/ Year	No. runs/ Week	No. runs/ Year	Calibrator, QC runs/ Year	Tot. Tests/ Year (pts+QC)
T4	624	3	156	572	1196
T3 Uptake	624	3	156	572	1196
TSH	8040	5	260	884	8924
T3	396	3	156	572	968
Free T4	5640	3	156	572	6212
HCG	2496	5	260	884	3380
Prolactin	720	3	156	572	1292
FSH	720	3	156	572	1292
LH	360	3	156	572	932
B12	600	3	156	572	1172
Folate	360	3	156	572	932
Ferritin	600	3	156	572	1172
Digoxin	1200	3	156	572	1772
Assumptions:					
Number of controls/run:		3			
Number of calibrators/week:		2			
Number of weeks/year:		52			

Table 2 calculates the annual costs and contribution for Analyzer A (excluding the sales rep's $1.00 incentive offer).

Table 3 does the same thing for Analyzer B and assumes that the tests marked with an asterisk will continue being performed by RIA.

Table 2.

Costs & Contribution on Tests for Analyzer A

Analyzer A--5 Year Lease, Lease Cost Included in Reagent Cost											
Test Name	No. tests/ Year	Material cost/test	Labor cost/test	Tot. direct cost/test	Material cost/year	Labor cost/year	Tot. direct cost/year	Price/ Test	Adjusted Rev. (uncoll. @ 26%)	Annual Contribution	Margin
T4	1196	$5.65	$0.83	$6.48	$6,757.40	$992.68	$7,750.08	$24.50	$21,683.48	$13,933.40	64.3%
T3 Uptake	1196	$5.65	$0.83	$6.48	$6,757.40	$992.68	$7,750.08	$24.50	$21,683.48	$13,933.40	64.3%
TSH	8924	$5.65	$0.83	$6.48	$50,420.60	$7,406.92	$57,827.52	$60.00	$396,225.60	$338,398.08	85.4%
T3	968	$5.65	$0.83	$6.48	$5,469.20	$803.44	$6,272.64	$55.00	$39,397.60	$33,124.96	84.1%
Free T4	6212	$5.65	$0.83	$6.48	$35,097.80	$5,155.96	$40,253.76	$48.00	$220,650.24	$180,396.48	81.8%
HCG	3380	$5.65	$0.83	$6.48	$19,097.00	$2,805.40	$21,902.40	$35.00	$87,542.00	$65,639.60	75.0%
Prolactin	1292	$5.65	$0.83	$6.48	$7,299.80	$1,072.36	$8,372.16	$74.80	$71,514.78	$63,142.62	88.3%
FSH	1292	$5.65	$0.83	$6.48	$7,299.80	$1,072.36	$8,372.16	$53.60	$51,245.89	$42,873.73	83.7%
LH	932	$5.65	$0.83	$6.48	$5,265.80	$773.56	$6,039.36	$53.60	$36,966.85	$30,927.49	83.7%
B12	1172	$5.65	$0.83	$6.48	$6,621.80	$972.76	$7,594.56	$51.25	$44,448.10	$36,853.54	82.9%
Folate	932	$5.65	$0.83	$6.48	$5,265.80	$773.56	$6,039.36	$55.00	$37,932.40	$31,893.04	84.1%
Ferritin	1172	$5.65	$0.83	$6.48	$6,621.80	$972.76	$7,594.56	$47.00	$40,762.16	$33,167.60	81.4%
Digoxin	1772	$5.65	$0.83	$6.48	$10,011.80	$1,470.76	$11,482.56	$53.00	$69,497.84	$58,015.28	83.5%

Table 3.

Costs & Contribution on Tests for Analyzer B

Analyzer B-Purchase; Continue Performing Some Tests By Current Method											
Test Name	No. tests/ Year	Material cost/test	Labor cost/test	Tot. direct cost/test	Material cost/year	Labor cost/year	Tot. direct cost/year	Price/ Test	Adjusted Rev. (uncoll. @ 26?	Annual Contribution	Margin
*T4	1196	$2.02	$2.48	$4.50	$2,415.92	$2,966.08	$5,382.00	$24.50	$21,683.48	$16,301.48	75.2%
*T3 Uptake	1196	$2.02	$2.48	$4.50	$2,415.92	$2,966.08	$5,382.00	$24.50	$21,683.48	$16,301.48	75.2%
TSH	8924	$2.86	$0.65	$3.51	$25,522.64	$5,815.47	$31,338.11	$60.00	$396,225.60	$364,887.49	92.1%
*T3	968	$1.97	$2.78	$4.75	$1,906.96	$2,691.04	$4,598.00	$55.00	$39,397.60	$34,799.60	88.3%
Free T4	6212	$2.86	$0.65	$3.51	$17,766.32	$4,048.15	$21,814.47	$48.00	$220,650.24	$198,835.77	90.1%
HCG	3380	$3.26	$0.65	$3.91	$11,018.80	$2,202.63	$13,221.43	$35.00	$87,542.00	$74,320.57	84.9%
Prolactin	1292	$3.26	$0.65	$3.91	$4,211.92	$841.95	$5,053.87	$74.80	$71,514.78	$66,460.91	92.9%
FSH	1292	$3.26	$0.65	$3.91	$4,211.92	$841.95	$5,053.87	$53.60	$51,245.89	$46,192.01	90.1%
LH	932	$3.26	$0.65	$3.91	$3,038.32	$607.35	$3,645.67	$53.60	$36,966.85	$33,321.17	90.1%
B12	1172	$3.26	$0.65	$3.91	$3,820.72	$763.75	$4,584.47	$51.25	$44,448.10	$39,863.63	89.7%
Folate	932	$3.26	$0.65	$3.91	$3,038.32	$607.35	$3,645.67	$55.00	$37,932.40	$34,286.73	90.4%
Ferritin	1172	$3.26	$0.65	$3.91	$3,820.72	$763.75	$4,584.47	$47.00	$40,762.16	$36,177.69	88.8%
Digoxin	1772	$2.86	$0.65	$3.51	$5,067.92	$1,154.75	$6,222.67	$53.00	$69,497.84	$63,275.17	91.0%

*Tests that will continue being performed by current method

Table 4 is the net present value analysis for Analyzer A. The Laboratory Manager made some assumptions about decreasing revenue and increasing labor rates, and plugged all of them into the NPV model. The NPV is $3,463,760, a very good return on an acquisition. (The NPV must be positive for a project to be considered good; the higher the number, the better the project). Now Company A is starting to look pretty good: no upfront costs, 6 months of essentially free reagents, and the opportunity to get away from RIA, even if Analyzer A is a little more labor intensive than Analyzer B.

Of course, the laboratory manager does the same analysis for Analyzer B. Despite the seemingly good offers and benefits from Company A, Analyzer B has a slightly better NPV Table 5. Thus, from strictly a financial viewpoint, the hospital would be better off purchasing Analyzer B than leasing Analyzer A.

Part of the net present value calculation includes a discount factor. The discount factor (also called a discount rate) should reflect the rate of return that could be obtained in the financial market on securities of similar risk to the acquisition.[15] Your own Finance Department should have a rate for you to use; talk to your financial manager to obtain your institution's discount rate.

Once you have determined cash flows and have your discount factor, you are ready to perform your NPV. If you have set up your spreadsheets similar to those in our example, calculating the NPV is as simple as using the spreadsheet's formula for NPV. In two of the more widely-used spreadsheets, the argument is @NPV(discount rate, range of cash flows from year 1 through year 5) plus the outlay in year 0.

In our example, then, we enter the formula into a cell and plug in the required information. The value which calculates is the NPV. If the number is positive, the acquisition is indeed worth more to the institution than it costs, and it can be considered a good investment. If the number is negative, the acquisition costs more than it is worth, and is not a financially sound pursuit.

Table 4.
NPV's for Acquisition Options: Cash Flows
Option 1: Lease Analyzer A; assume ownership at end of 5 years

	Initial Cash Flow (Year 0)	CF Year 1	CF Year 2	CF Year 3	CF Year 4	CF Year 5
Adjusted revenues		$1,139,550	$1,116,759	$1,094,424	$1,072,536	$1,051,085
Initial cash flow	$0					
Cost of lease		$0	$0	$0	$0	$0
Cost of materials		($85,992)	($171,986)	($171,986)	($171,986)	($171,986)
Cost of labor		($25,265)	($26,023)	($26,804)	($27,608)	($28,436)
Cost of service		$0	($10,000)	($10,000)	($10,000)	($10,000)
Travel for training costs		($4,800)	$0	$0	($2,400)	$0
Computer interface costs	($10,000)	$0	$0	$0	$0	$0
Salvage value at end of lease		$0	$0	$0	$0	$11,500
CASH FLOW	$0	$1,023,493	$908,750	$885,634	$860,542	$852,163
				NPV:	**$3,463,760**	

Assumptions:
Revenue decreases 2% annually
Cost of labor increases by 3% annually
Annual service is negotiated with vendor at $10,000
Company supplies 6 months of materials and reagents for $1.00

Table 5.

NPV's for Acquisition Options: Cash Flows

Option 2: Buy Analyzer B and Continue with Current Methodology for Some Tests

Initial Cash Flow	Initial Cash Flow (Year 0)	CF Year 1	CF Year 2	CF Year 3	CF Year 4	CF Year 5
Adjusted revenues	$0	$1,139,550	$1,116,759	$1,094,424	$1,072,536	$1,051,085
Initial cash flow--purchase Analyzer B	($85,800)					
Initial cash flow--cost of starter kit	($1,500)					
Cost of materials	$0	($88,256)	($91,787)	($95,458)	($99,276)	($103,248)
Cost of labor		($26,270)	($27,058)	($27,870)	($28,706)	($29,567)
Cost of service		$0	($10,000)	($10,000)	($10,000)	($10,000)
Travel for training costs		($2,500)	$0	$0	$0	$0
Computer interface costs	($8,000)	$0	$0	$0	$0	$0
Salvage value at end of five years		$0	$0	$0	$0	$5,500
CASH FLOW	($95,300)	$1,022,524	$987,914	$961,096	$934,553	$913,770
				NPV:	**$3,578,502**	

Assumptions:
Revenue decreases 2% annually
Cost of labor increases by 3% annually
Annual service is 10% of instrument's list price
Cost of materials increases 4% annually

IV. REDUCING LABOR COSTS

FTE REDUCTION

One of the obvious ways to reduce the cost of labor in the Chemistry Laboratory is to reduce the number of employees. While this sounds simple, it is a factor that is frequently overlooked--because we are human and develop close ties to those with whom we work.

A good supervisor is one that develops a spirit of teamwork within the work group itself and between the supervisor and the group. Such a mindset doesn't make a reduction of the number of members in the group one of the first ideas that comes to that supervisor's mind as an attractive solution to cost-effectiveness. It may take someone from outside the department or outside of the laboratory to mandate such a change.

For the past five years in our hospital we have had a yearly hospital-wide project to decrease costs called Operation Improvement (O.I.). In the O.I. process, goals are set for reduction of costs for different departments or combinations of departments. Work groups are formed using supervisors/managers/administrators from the departments and, for at least three months, the achievement of cost reduction is made their highest priority.

The groups solicit ideas from the employees in their departments. For example, each employee may be asked to submit at least five ideas for decreasing costs. To implement the ideas that are developed, the hospital provides funding for up to three times the amount of the savings. Many good ideas that have nothing to do with reduction of personnel have been developed using this method, but it is from this process that mandates for reduction of personnel have come. Further, the Chemistry Laboratory has benefited in that it has obtained very modern equipment, thanks to ideas brought forward in the O.I. process.

Fortunately, in most cases in our chemistry laboratory, FTE reductions have been made less painful by using attrition or by allowing/encouraging personnel to work part-time on a permanent basis. We had previously discouraged part-time employment so we have many employees who are happy to work less than a full-time schedule. When such reductions are not sufficient, then we have imposed mandatory time-off (flex time) during times when there were minimal numbers of employees on vacation or when the workload of the department was decreased for whatever reason, e.g. fewer surgeries done during the holiday period, etc.

We have tried to make this process as painless, or at least as fair, as possible by asking for volunteers first, and then, when there are no volunteers, keeping good records of the non-voluntary time-off imposed on each employee.

There are, of course some negative aspects to this approach. Having more employees work a part-time schedule but still receive benefits produces increased costs for the hospital in benefit expenses, as compared to actual FTE reductions. Scheduling is not as easy as it previously was, because we have people who work only three days a week, four days a week, nine days out of ten, etc. But we have been able to maintain morale at a reasonable level by working with employees to eliminate over-staffing rather than by just imposing lay-offs.

This approach makes it easier to avoid over-time costs because when workload increases, FTE hours required to complete the workload increase, and employees then use hours which they would otherwise have off (either because they are part-time or because they are required to take flex time). It is much easier to find staffing for a temporary project because the part-time employees are often willing to work some extra days on a temporary basis.

TEST METHODOLOGY

Another way in which labor costs can be decreased is through the choice of test methods. Obviously, automated methods should be employed whenever they are comparable in accuracy and precision to non-automated methods. It is especially important to try to automate methods that must be done on a 24-hour basis so that call back of employees from other shifts can be avoided on the evening and midnight shifts. Having been a chemistry supervisor before automated methods for ammonias were available, Evans can relate to the importance of avoiding call back of employees.

All too often she received a call from the laboratory director in the middle of the night saying that he had approved a stat ammonia and needed to find someone to come back to the lab to do it. This definitely was neither fun nor cost-effective.

The choice of methods can be especially important for very long procedures such as toxicology screens and isoenzymes by electrophoresis. Reliable methods are available which may not completely replace the tedious time-consuming methods, but which can decrease the time spent on those methods.

CK MBs by automated methods are one example of such a time-saver. They are subject to false positives but not to false negatives, so that patients in the E.D. or Coronary Care Unit can be sent home much faster if these methods are run on a stat basis and the results are negative.

Urine drug screening by quick immunoassay using monoclonal antibodies is another example. When these screening assays are positive, then the time-consuming toxicology screens must be run if it is necessary to determine exactly which drug of abuse is present. But if the results are negative, many hours of laboratory time can be saved as well as many hours of costly Emergency Department time.

INSTRUMENT SELECTION

The choice of instrumentation also has a direct impact on labor costs. Automated instruments that require frequent calibration and lengthy preventive maintenance add to labor costs as well as reagent costs. But of even greater importance is the reliability of the instrumentation and the resulting minimization of down-time and trouble-shooting, especially if the available back-up procedures are non-automated and time consuming.

TRAINING

Another factor which is involved in the cost of labor for the chemistry department is the cost of training new employees. In our chemistry department there are thirteen work areas that must be staffed. When new employees are hired, we train them by having them spend at least a week of orientation in the general department, including the laboratory information system, Quality Control, and Quality Assurance. During this week the technologist who is doing the training does little or no productive work as far as the department workload is concerned.

After the orientation period, training in the stat area begins. The newcomer spends one week with another technologist learning each stat position. In most cases, during the next week they are able to work the position alone, but they often require some support from other techs. Essentially during this time they are working as a productive FTE only half of the time.

When rotation outside of the stat area occurs, often more than one week of training is required to learn the positions since they are less automated and more labor-intensive.

Determining how many techs to train for each position is a difficult decision. In a larger laboratory, it may not be cost effective to keep everyone trained in all positions because in some of the more esoteric positions, such as toxicology or the estrogen receptor position, technologists must rotate through the area fairly frequently in order to maintain their skills in those areas. If everyone were trained in those positions, we wouldn't be able to rotate people through the positions frequently enough for maintenance of skills, and additional re-training costs would be incurred. On the other hand, it is necessary to keep enough people trained so that the positions can be covered when people are on vacations and emergencies occur, such as illnesses etc.

An additional consideration is maintaining some minimal degree of sanity for the technologists who rotate through areas that provoke higher stress, such as the stat area. Trying to find the right balance in the number of people trained for each area is very challenging. An additional factor is that some techs would prefer to rotate to a different position each week, while some techs would prefer to stay in the same area forever. While it is not necessarily the supervisor's responsibility to keep techs entertained, a supervisor should make every effort to keep the job interesting and challenging for his or her employees, and rotating them in areas they most enjoy is one way to do it. It is important to maintain morale or eventually valued employees leave, and the supervisor faces additional training costs.

Labor costs can be contained by reducing turnover (employee retention) and minimizing training costs. The next sections address ways to achieve those goals.

COST CONTAINMENT THROUGH EMPLOYEE RETENTION

Interview Techniques: Hiring the right person when an opening occurs is one of the most important projects that a supervisor undertakes. It is vital that a supervisor takes the time to prepare for an interview. A process that works very well is the group interview process. This technique, while very valuable in making a good selection, is obviously feasible only in areas of the country where the applicant pool is comprised of sufficient numbers of qualified personnel. In this era of personnel shortages, many laboratories do not have the luxury of selecting from many qualified applicants.

In group interviews, a standard set of questions is asked by each of the interviewers. Evaluations of the applicant's responses are written on a standard form by each member of the interviewing staff. Identical questions are used to interview at least three candidates for each position. When the interviewing is complete, the evaluators meet to discuss the responses of each applicant. Often, there is at least one candidate with whom everyone is comfortable. If consensus cannot be reached, Human Resources is asked to forward additional resumes, and the process is repeated.

Start by choosing an interview team that has from three to five members. The team obviously includes the prospective supervisor(s), but it can also include other levels of management such as a chief tech, lab manager, etc. When interviewing for supervisory or lead positions, it is a good idea to include interviewers who will become colleagues of the applicant, since they will undoubtedly work together very closely.

When the selection process is complete for choosing the interview team, the members meet to discuss which questions will be asked and by whom. The team prepares an evaluation form on which all members agree.

Then do some role-playing; practice with a volunteer (or commandeered) "applicant." One caveat about role-playing: it may not be a comfortable situation for either the interviewer or the mock applicant, but the practice makes the interviewer more familiar with the questions--he or she will have the opportunity to improve probing techniques and to ask questions clearly. The interviewer will learn which questions are good ones and which need improvement. It is a staged situation and will feel like a staged situation (though over time with practice, most people become more comfortable with role-playing).

The choice of questions that the team uses should be made carefully, and follow-up secondary or tertiary probes should be prepared to be used if the response received is vague or ambiguous. Here are some examples:

1. "All of us have some things that we are really good at and some things that we need to work on. Can you give us an example of something that you need to work on?"

This is a common interview question and the applicant's response can give the interviewer some idea of the preparation that the applicant put into the interview. If the applicant seems unprepared for the question, it may say something about the applicant's ability to plan and anticipate. If an applicant has a hard time or can't answer this question, it might indicate that he or she is not a quick thinker, or doesn't have sufficient self-esteem or confidence to share something about him/herself that needs improvement, or that he or she is not terribly honest or cognizant of his or her own abilities.

2. " What was something about your last job that you really didn't like?"

If the response is "Doing heel sticks on babies." and the position in question involves drawing blood in the nursery, then the applicant may not be the best fit for the job (although nearly all jobs have some undesirable tasks to them).

If the response is something like "My job was repetitious and wasn't challenging" then its time for the secondary probe "What exactly did you do?" or "What would have made the job challenging for you?" Perhaps this person is more suited for a research position on the cutting edge of technology, since routine lab work is frequently repetitious.

If the response is "My boss was terrible", and after the secondary probe, the applicant goes on to describe you or the laboratory director, perhaps this is not the right person!

3. "Tell us about a laboratory problem that you solved recently and that you are proud of the way you handled the situation." The response to this question can provide insight into the applicant's initiative and trouble-shooting/ problem-solving abilities. It may also give you an indication about his/her judgment in knowing which problems should be

handled at the technologist level and which should be referred to a supervisor. It is always interesting to hear if they mention documentation of the problem and communication to a supervisor at a later time.

4. "We are interviewing a number of applicants for this position. Tell us why you think you should be our first choice."

For many people, this seems to be a difficult question to answer; many people feel that they are bragging and are concerned about appearing conceited or arrogant when they are asked to sell themselves. The response may give you insight about what qualities this person considers important to doing a good job and may say something about the applicant's self-confidence.

The interview nearly always makes the applicant feel some degree of stress; that is not to be avoided. Most positions in laboratories are stressful at least part of the time; this is the interviewer's opportunity to see how the applicant deals with a stressful situation. If there is an awkward silence when he or she can't think of a response to a question, do not help! Remember, the object is to obtain the applicant's answer, not the interviewer's. The interviewer can prove that he or she is a really a nice person later, at the end of the interview, but not during the questioning section.

The above are examples of questions that can be asked in an interview. It is also important to know what questions cannot be asked during the interview process. Milkovich and Glueck, in *Personnel/Human Resource Management: A Diagnostic Approach*[16] list the following questions as those to avoid:

1. Age - You can't ask, "How old are you" or "Please give us an estimate of your age". You can ask, "Are you over 18?"

2. Sex - You cannot ask what a person's sex is, or even make notes or comments regarding sex, unless it is a bona fide occupational qualification.

3. Physical abilities - You cannot ask how tall the applicant is or how much he/she weighs, but you can describe the type of manual labor or lifting requirements of the job. You can require a physical exam, which may help you obtain the information, if it is relevant to the position.

4. Marital status - You cannot ask whether a person is married, single, engaged, divorced, or living with someone.

5. Children - You cannot ask whether the applicant has children, or how many, or whether he or she plans to have children. If they volunteer the information that they do have children during the interview process, you can't ask how they plan to care for them while they are at work.

6. Housing - You can't ask if they own or rent a home or whether they live in an apartment or a house, etc.

7. Military status - You can ask if the applicant is a veteran but you cannot ask what branch of the service they served in, or what type of discharge they received.

8. Criminal record - You can't ask the applicant if he/she has ever been arrested, convicted, or spent any time in jail.

Before the interviewing process, each member of the team should have copies of the job applications and resumes, and should read them and then bring them to the interview. Look for any substantial time gaps between jobs or schooling. It is appropriate to ask what happened during that time.

The application should have a list of the instruments used in your laboratory and a check-off section where the applicant can indicate experience with the particular instruments. If the application does not include such information, one of the interviewers should use a standard form developed in the laboratory to gather the information. If the applicant indicates that he or she is familiar with an instrument that the laboratory uses, a supervisor or technologist can ask some very specific troubleshooting questions, which can help the interview team determine the applicant's technical expertise.

Looking at starting and ending salaries and the time between promotions can give you an indication of the quality of their performance.

Below is an example of a interview evaluation form that can be used. You will want to modify it to fit your laboratory situation and values.

	SUPERIOR	VERY GOOD	GOOD	FAIR	POOR
JOB KNOWLEDGE AND SKILLS					
EDUCATION					
EXPERIENCE					
ORGANIZATIONAL SKILLS					
COMMUNICATION SKILLS					
POISE AND BEARING					
MANNER AND ATTITUDE					
INITIATIVE/RESOURCEFULNESS					
MOTIVATION/AMBITION					
STABILITY					
LEADERSHIP ABILITY					

APPLICANT'S STRONG POINTS______________________________

APPLICANT'S WEAK POINTS______________________________

COMMENTS______________________________

This interview and selection process does take an investment in time, but it is small compared to the time that will be spent in counseling and documenting poor performance when a bad hiring decision is made. Even when the best decision-making tools are used, occasionally wrong choices are still made. It is very important to carefully observe and evaluate performance during a new employee's probationary period.

Early identification of poor performance is not always a straightforward, easy call. Every new employee has a learning curve and a good supervisor wants to be fair and give a new employee a chance, but if he or she cannot adapt to the laboratory situation, it is much easier to correct the error early on.

Participation in a Medical Technology training program is a good way to assure that your laboratory has qualified applicants for any vacancies that may occur. Not only do you have a chance to train the participants in what you perceive to be the right techniques and work habits, but you have the opportunity to observe them for six to nine months and to learn if their personal "opportunities for improvement" are those that you can tolerate in your lab.

Reducing Turnover by Keeping Morale High

Reducing turnover by maintaining a high level of morale is another challenge every manager and supervisor faces. While morale can be affected by an almost infinite number of factors, there are a number of ways a manager can provide enriching and rewarding jobs for their employees. Following are a few ideas that an effective leader can implement.

Staff Participation in Development Projects: It is very tempting to hire or assign one technologist, perhaps with an advanced degree, to do all of the projects for checkout and adaptation of new instruments and/or methods for a laboratory. It may make the jobs of supervisors and clinical chemists easier to some degree and it is a very interesting and satisfying job for the one person assigned to that position, but it denies all of the other technologists a chance to do any of the non-routine and challenging projects for the department.

In most situations a better approach is to allow all of the technologists who are interested and capable, the opportunity to work on development projects. In our chemistry department, as each new project comes up, the supervisor discusses it with different technologists who may have a particular interest in a specific area (or who simply may be interested in a new challenge). As you well know, medical technologists by nature are intelligent and inquisitive; most of them will welcome the chance to participate in a project that offers them a learning experience and a break in the routine.

The technologist selected for the project is usually selected from the group of techs who frequently work in the area where the method will be done. For very large projects in our laboratory, such as the development of the two "work-horse" automated instruments, two technologists have worked on the project as a team.

The techs work closely with the supervisor and clinical chemist to be sure all of the technical aspects of checking accuracy, precision, linear limits, normals, interferences, etc. are done thoroughly and well. They write the methods which are then approved (or revised and approved) by the supervisor/chemist. They also address the challenge of how to make the instrument/method work smoothly within the particular area, since they are uniquely qualified to assess what is best for their section. They design check-off lists and schedules for recording of maintenance, work lists, and order sheets for maintaining inventory. They organize the work area itself to maximize efficiency in running the tests, performing routine maintenance, and facilitating troubleshooting.

Because of their work experience and familiarity with the area and its interactions with the rest of Chemistry as well as the Specimen Processing Department, they are able to do a better job than most full-time Research and Development technologists could do. But even more importantly, morale is improved because they have an occasional chance to do something non-routine, and that allows them to utilize the knowledge for which their years of college and job experience have prepared them.

Educational Opportunities: Another way to maintain morale in the clinical chemistry laboratory is to promote educational opportunities. It is important to encourage the employees to pursue as many continuing education (CED) activities and seminars as reasonably possible. Some laboratories are able to provide financial assistance for educational activities, however, cutbacks in funding have frequently limited the degree to which institutions can finance outside educational programs. That does not mean that continuing education should not be encouraged.

Our laboratory has a Laboratory Continuing Education Committee, co-chaired by the Administrative Services Manager and a Clinical Chemist, that has taken advantage of our in-house speakers on a wide range of topics that are of interest and have practical value in widening the scope of knowledge of our staff.

Some examples are lectures by physicians in various specialties such as bone marrow and liver transplantation and the activities of the Stroke Center. With the help of our Oncology Nurse Manager, we recruited a speaker who was a patient celebrating his one-year anniversary as a bone marrow transplant. In one of our most captivating CED lectures, he shared his experience with us by giving us the patient's perspective of the procedure and of the post-transplant period.

Another extremely interesting lecture on Robotics was given by an engineer in our Clinical Engineering Department. Our lab has videotaped these lectures and the tapes are available to be viewed in the lab or to be checked out and viewed at home for those unable to attend the lectures.

Although a laboratory in a smaller institution may not have access to quaternary care specialists or videotaping capabilities, the specialists who practice in smaller institutions undoubtedly have interesting cases to discuss in which the laboratory's involvement made a major difference in a patient's outcome. They may also have a particular interest in an area of their specialty on which they could lecture to the laboratory. Because these physicians are helping educate the staff members of the laboratory that they most frequently use, they rarely expect or accept fees or honorariums.

Professional organizations nearly always include as part of membership some sort of CED opportunities. Some of these are self-study programs available at a relatively inexpensive rate to members. Other organizations may offer membership in local chapters that provide educational lectures for a very nominal fee (or free to members). Laboratory professionals should take full advantage of such educational opportunities in their own areas.

Reducing Retention Through Effective Leadership

Dr. James J. Gooch, Director of Biotechnology Systems for Dupont, recently led a seminar on *"Strategies for Achieving Organizational and Personal Excellence"*[17] in which he described ways in which the workplace environment and management concepts are changing. According to Dr. Gooch, organizations are becoming flatter with fewer layers of management; there is more diversity within work groups, and the prevailing use of Continuous Improvement Initiatives is improving total quality, productivity, profit, and employee satisfaction. The true essence of leadership is the ability to influence and inspire others. Leadership in an organization must :

1. Establish a real vision for the future;
2. Define organizational values;
3. Accomplish the needed changes by:
   a. Effective communication and motivation;
   b. Unleashing the full potential of its employees.

The success of an organization or business depends on

the ability of its leaders to formulate a clear mental picture of its mission, to create a vision that is superior to that which others have created, and to communicate and transfer that vision to others. Employees must individually and collectively buy into the vision and must embrace the leader's vision as their vision; only when this has occurred will the employees accept the responsibility for achieving that vision.

Dr. Gooch advocates managing by principles rather than creating rules and policies; i.e. the effective manager sets boundaries and parameters and then holds the employees accountable for the results. He believes that leaders should not *force* change but should *facilitate* change and involve employees in the process of effecting the change process and in being the agents of creating change.

The premises of this management approach are:

1. The business/organization is actually the people who comprise it and the success of the business depends on the people who create it.
2. There is a clear relationship between how people relate to the organization/business and what the business delivers.
3. Getting skilled people to approach their role in the business with the spirit and will to enable the business to achieve its vision is largely the process of leading.

Leadership must recognize that people want:

1. Respect as an individual.
2 More responsibility.
3. More control and ownership of their work.
4. An increased sense of contribution.
5. A voice in the process.
6 Confirmation that what they say and do does have an impact.
7. Opportunities to grow and develop their full potential.

The barriers to the delegation of authority and responsibility for supervisors and managers are: fear of the unknown, fear of a loss of control, fear that mistakes will be made, and even fear that a good job will be done ("if someone else does a good job, then what will my role be?") Managers must manage their desire to have things done only their way. They should surround themselves with successful people and then get out of their way.

Finally, Dr. Gooch proposes the following framework for empowering others:

1. Define the strategic direction.
2. Ensure alignments:
   Functional and personal objectives;
   Operating principles (not rules);
   Support systems and structures in place.
3. Define the boundaries:
   Responsibility
   Authority
   Accountability
   Performance expectations
   Levels of risk
   Consequences
4. Provide the resources:
   Knowledge and information
   Training and development
   Tools, materials, and facilities
   Identification of resource providers
5. Be supportive:
   Communicate openly
   Give respect, trust and confidence
6. Define the auditing process:
   Required interactions
   Meters and indicators
   Accountability for behaviors and actions.
   Variances and constructive feedback.
7. Manage by results:
   Leave people alone and let them do the work.

V. REDUCING OVERHEAD COSTS

Some of the most difficult costs for a chemistry supervisor to control are those classified as overhead expense, since those costs are usually out of the supervisor's control. A well-known edict for stress reduction is "*do not concern yourself with things you cannot control.*" However, there are actions which can be taken to reduce overhead, especially the direct overhead costs found within the laboratory (very little can be done at the supervisory level to control institutional overhead). In the event that a chemistry supervisor handles stress well and can afford to be concerned with things he or she cannot control, the supervisor may offer suggestions to his or her manager or director that can bring overhead costs down. Some ideas that may be within a chemistry supervisor's scope of influence and would help in reducing overhead costs are included below.

CENTRAL PROCESSING/ACCESSIONING

Most modern, large chemistry instruments can accommodate primary tube sampling, in which the actual evacuated tube used for phlebotomy is centrifuged and put directly on the instrument, without aliquotting. If your chemistry instruments have this capability, its use is something that you should seriously consider. The reduction in aliquotting time could result in significant labor savings in the Processing Area. While aliquotting will probably never be completely eliminated, holding it to a minimum is a cost-effective move that a chemistry supervisor *can* influence.

If your phlebotomists do not use evacuated tubes with gel barriers, then extra time is being spent pipetting aliquots or inserting serum separators instead of pouring directly from the barrier. Again, while there are exceptions, the majority of chemistry tests can be successfully tested on serum collected in a gel barrier tube, and the chemistry supervisor and clinical chemist should authorize their use whenever practical.

INCREASED AUTOMATION/ADVANCES IN TECHNOLOGY

Reducing labor in specimen processing is currently under intense scrutiny in the laboratory market. Many innovative ideas have been developed into working models. These include barcoding the patient's armbands for positive identification, barcoding specimens, and using aliquot labels (especially effective when used on instruments with bidirectional interfaces). Also linking remote testing sites to the LIS, allowing technologists to review results of tests performed in a nursing unit, using robots which travel around the hospital, employing robotics within the Specimen Processing Area, and instituting reliable pneumatic tube systems, conveyor belts, and specimen transport "cars" to deliver specimens throughout the laboratory reduce labor costs. Many of these ideas require significant capital outlays, but can result in a sizable return on the investment. The re-engineered laboratory of the future will no doubt incorporate these and may other step-saving ideas; the chemistry supervisor should support the implementation of these ideas as the budget permits.

A number of laboratories have combined the highly automated tests from different sections into one automated laboratory. In these models, the technologists might run the automated cell counter traditionally found in Hematology and also run the workhorse chemistry analyzer. Automated coagulation analyzers, urinalysis instruments, and immunoassay instruments may also be installed into an automated section. Not only has this idea been used successfully in many institutions, but many technologists like the change and the opportunity to use different skills in which they were trained. A chemistry supervisor could approach upper management with this concept, and participate in the planning and development of a successful automated laboratory.

While earlier in this chapter we provided a case study that supported centralization of laboratory testing, decentralized, point-of-care testing is not going to disappear soon. The reasons for this are many: manufacturers of point-of-care test systems are relentless in their marketing efforts to physicians, nurses, physician assistants, and others. They capitalize on the frustration that these professionals experience in waiting for results from the laboratory; after all, why wait for 30 minutes or more when professionals close to patients can simply, quickly, and reliably obtain results themselves in fewer then 5 minutes?

If your laboratory is not prepared to undertake the steps needed to keep testing centralized in the laboratory, you can best benefit the patient by becoming an ally of the doctor or nurse who is testing at the bedside. As described in our earlier example, we found nurses to be especially thankful for expert help in any laboratory testing. In most cases, CLIA has taught them that there is much more to achieving a good lab result than applying blood or urine to a reagent strip. Cooperate and support them in their efforts to establish an effective decentralized testing program--help them understand the importance of consistent documentation, running controls, using reagents and controls that have not expired, storing reagents and controls properly, performing regular preventive maintenance of instruments, and solving interferences that they may encounter.

The laboratorian is the expert on CLIA--no one has had to digest it as much as the laboratory director, manager, supervisor, and technologist. Share your expertise with nurses and physicians. Help them stay within the guidelines that CLIA prescribes; assist them in recognizing what they can and cannot do by law. Your approach must be professional and warm, and your words and actions must convey that you are there to help them--many doctors and nurses believe that the laboratory's attitude will be defensive and territorial (i.e., that the laboratory will not want to help them "take business away"). Even if it is difficult for you or another laboratory professional to contain your resentment of others encroaching in the laboratory's domain, you must get past your anger. Bedside testing is most likely here to stay, and being a team player will help non-laboratorians to provide good testing, that benefits the patient, while reducing the number of costly errors that they may otherwise make without your help. Besides, as was said before, it is a time waster to concern yourself with things you cannot control.

VI. SUMMARY

While this chapter addressed cost containment directed specifically toward clinical chemistry, many of the principles are just as germane in a cross-functional environment.

The most recent experience of the authors has been in a large, university hospital setting; however, they have also had laboratory management experience in a smaller hospital setting and clinic laboratories. By definition, university and teaching hospitals face more difficult challenges in containing costs than do community hospitals or HMO hospitals.

In the last several years, upper management in all institutions has placed increasing emphasis on cost control, to the point that most are now requiring a look at the laboratory from a make versus buy standpoint. One scenario of what the future may hold for hospital (especially small to medium sized) laboratories is one of outsourcing nearly all laboratory services. As the profit margin for laboratory testing disappears, it may be more cost effective for hospitals to invite a reference laboratory in to direct on-site whatever minimal stat testing is necessary for managing an acutely ill patient. This would be a very small menu of tests, indeed, comparable to a "stat list" that many laboratories have assembled to improve turnaround times. All other testing, that which is not required immediately for managing a patient, would be transported to the reference laboratory, which would have a CPU-CPU link and look essentially like a laboratory down the hall rather than one remotely located. Overnight turnaround time would provide everything the physician needs for the care of his or her patients.

While various reference laboratories have attempted such models, most arrangements are not yet perfect in their goal, but soon will be. Even now, many reference laboratories offer such competitive pricing that in some cases, the cost of buying the test from a reference laboratory is cheaper than the cost of performing it in-house. While there are always exceptions to the rule, the large, well-known reference laboratories offer high quality testing and good service, making it increasingly difficult for laboratory administration to justify the high cost of in-house testing when the same test can be purchased for less. The only recourse that a hospital laboratory has is to reduce its costs to the lowest it possibly can, and to implement the most efficient test systems available. Only then can the hospital or other healthcare facility laboratory compete and remain viable in the future.

VII. INDEX - HCFA APPROVED PROFICIENCY TESTING PROGRAMS

Accutest
800-356-6788

American Academy of Family Physicians
800-274-2237

American Academy of Pediatrics
708-228-5005

American Association of Bioanalysts
800-234-5315

American Osteopathic Association
312-280-5800

American Association for Clinical Chemistry
800-892-1400

American Proficiency Institute
800-333-0958

American Society of Internal Medicine
800-338-2746

American Thoracic Society
212-315-8808

California Thoracic Society
714-730-1944

College of American Pathologists
800-323-4040

EXCEL Program, CAP
800-323-4040

Commonwealth of Pennsylvania

215-363-8500

NY Department of Health
518-474-8739

NJ Department of Health
609-530-6172

Idaho Department of Health
208-334-2235

Ohio Health Department
614-466-2278

Pacific Biometrics Research Foundation
206-233-9195

Puerto Rico Health Department
809-764-6945

Wisconsin State Laboratory of Hygiene
800-462-5261

VIII. REFERENCES

1. The Advisory Board Company, 1993: *Vision of the Future-Briefing for Board Members, Physicians, and Administration*, Washington, D.C.
2. Homgren, Charles T: *Introduction to Management Accounting*, 6th Edition, 1984, Prentice-Hall, Inc.
3. Ibid.
4. Kieso, Donald E., and Weygandt, Jerry J.: *Intermediate Accounting,* Fifth Edition, 1986, John Wiley & Sons, Inc.
5. *Clinical Laboratory Improvement Amendments of 1988; Final Rule.* Federal Register, Friday, February 28, 1992, Vol. 57, No. 40.
6. Jahn, M: A Precarious Balance on the Shaky Dollar, *MLO*, January 1993.
7. Milkovich, G. T., and Glueck, W. F.: *Personnel/Human Resource Management: A Diagnostic Approach,* Fourth Edition, 1985, Business Publications, Inc.
8. *Workload Recording Method & Personnel Management Manual,* 1992, Edition, CAP, Northfield, IL.
9. Ibid.
10. Elevitch, Franklin R., and Della Rocchetta, Virginia: Pricing a Capitated Laboratory Contract: 7 Steps to the Poor Farm ?, *MLO*, October 1994.
11. Nierenberg, Gerard I: *The Art of Negotiating,* Simon and Schuster, 1968, 1981.
12. Ibid.
13. Fisher, Roger, Ury, William, and Patton, Bruce: *Getting to Yes,* Second Edition, 1981, Penguin Books.
14. Albrecht, Karl and Albrecht, Steve: *Added Value Negotiating*, 1993, Richard D. Irwin, Inc.
15. Franks, Julian, Broyles, John E., and Carleton, Willard T: *Corporate Finance: Concepts and Applications,* 1985, Kent Publishing Company.
16. Milkovich, George T., and Glueck, William F.: *Personnel/Human Resource Management: A Diagnostic Approach,* 1985, Business Publications,

Inc.

17. *CAMLT* Summer Seminar, July 10, 1994.

Chaper II

Cost Containment In The Clinical Hematology Laboratory

Valerie L. Ng, Ph.D, M.D.

Carol Hogan, MBA, MT(ASCP)

Eugene L. Gottfried, M.D.

Department of Laboratory Medicine, University of California, San Francisco, and Clinical Laboratories, San Francisco General Hospital, San Francisco, CA

I. INTRODUCTION

IMMENSITY OF TESTING IN UNITED STATES

Expenditures for health care in the United States have been rising at an alarming rate over the past thirty years (Figure 1).[1] Following World War II, new technology and automation began to transform medical practice and clinical laboratories. The realization that health costs have been rising even more rapidly than the general cost of living has led to demands for cost containment and health care reform.

Fig. 1. Health Care Expenditures as a Percent of the Gross Domestic Product

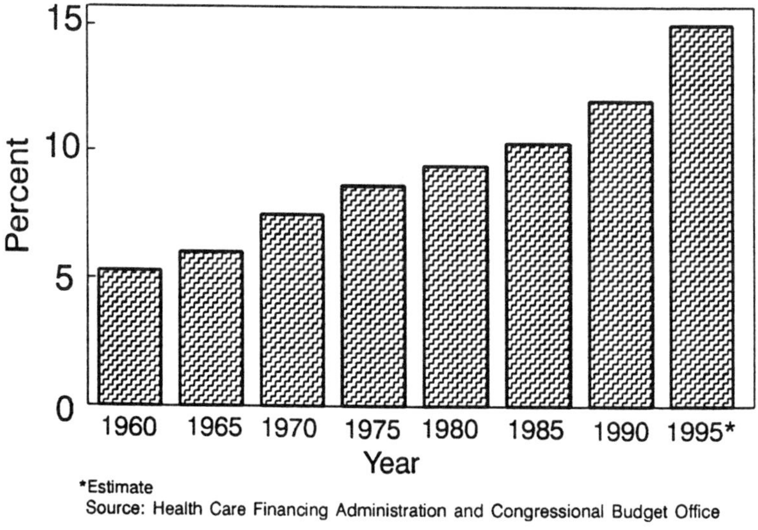

*Estimate
Source: Health Care Financing Administration and Congressional Budget Office

Laboratory testing increased dramatically after the first automated analyzers were introduced in the 1950's. In the typical laboratory, the workload followed a pattern remarkably similar to that of health care costs (Figure 2). More laboratory tests are being performed on patients today than a generation ago, no matter what measures one uses tests per outpatient encounter, tests per inpatient day, or (despite the trend toward shorter hospitalization) tests per hospital stay.

Fig. 2. Clinical Laboratory Workload *

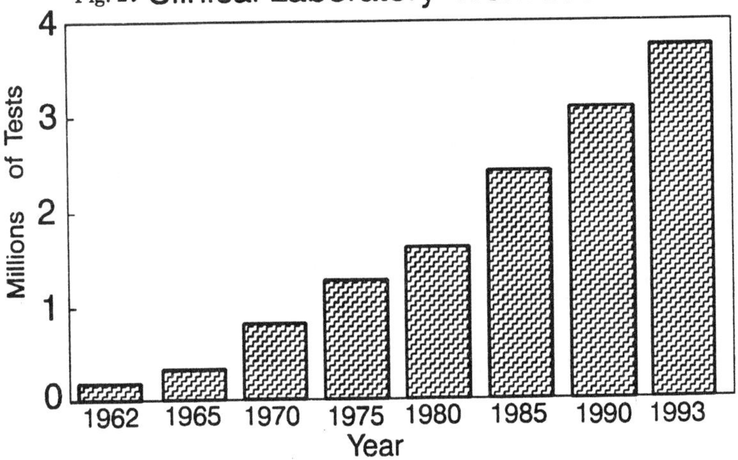

* Adjusted to reflect changes in the methods used to count test panels.

The costs of laboratory testing contribute substantially to the total cost of health care. Clinical laboratory charges in an acute-care hospital may constitute 15% to 25% of the hospital bill. In the United States, Medicare short-stay hospital charges for laboratory services increased from 11.6% of the total in 1983 to 15.3% in 1990.[2] In the Japanese and German national health plans, expenses directly attributable to laboratory testing constitute 2.5 and 6.5% of all medical costs, respectively.[3,4]

Many reasons have been proposed to explain the increasing volume of laboratory tests. Technology has played a key role.[5] Centralized hospital laboratories were developed in order to accommodate the automated analyzers that emerged in the mid-1950's. These instruments were typically bulky and expensive, but they could perform a large number of tests in a short time, at a lower cost per test than the manual methods they replaced. Earlier, many of the lab tests in hospitals were performed by nurses or interns in utility rooms or in small labs situated near each patient care unit. The test menu was generally very limited, and the knowledge that the one who ordered the procedure would probably have to do the analysis served as a powerful disincentive for unnecessary testing. The introduction of automated analyzers improved not only accuracy and precision, but also the variety of tests and analytical turnaround time. Physicians began to rely more on laboratory studies than on the traditional medical history and physical examination, which they perceived as "low-tech", less objective and less precise than a battery of laboratory tests.

With the current generation of automated hematology analyzers, for example, a complete blood count (CBC), including a leukocyte differential and platelet count, can now be performed in less than two minutes. This is a small fraction of the time formerly required for the same procedures to be done manually. As a consequence, doctors have learned to expect the results much sooner and repeat the tests more often. (This is an example of

Gottfried's Corollary to Parkinson's Law, "Expectations rise to meet new technology.")

New laboratory procedures are introduced at a faster rate than outdated methods are discarded, and the test menu continues to expand. Even when a new technique is intended to replace an obsolete assay, physicians are likely to request both the new test and the old one.

WHY WAS THIS TEST ORDERED?

Doctors order laboratory tests for a number of reasons (Table 1). Often, the diagnosis is already evident from the medical history and physical examination, and the management plan is clear. In such cases, the laboratory tests are ordered in an attempt to reduce the physician's (or patient's) level of uncertainty, or to assure "completeness."

Under prepaid managed care programs, there is increasing pressure to reduce costs and to resist the urge to order tests without a valid indication. There is growing recognition of the need to evaluate the clinical utility not only of newly introduced laboratory procedures, but also of tests already in common use and well established in current medical practice.

CRITERIA FOR A "GOOD" TEST

The traditional criteria for judging the value of a laboratory test focused on the technical and financial aspects of the procedure (Table 2). Unfortunately, a test may be fast, inexpensive and technically excellent, yet contribute little or nothing to the patient's care. Table 3 lists a new set of characteristics more appropriate to today's demands for clinical utility.

Table 1: Why Doctors Order Laboratory Tests

1. **Diagnosis**
   a. Screen apparently healthy individuals at risk for disease
   b. Exclude possible disease in selected patients
   c. Confirm probable disease in selected patients
2. **Management**
   a. Follow the course of disease
   b. Monitor the effects of treatment
   c. Determine the patient's prognosis
3. **Completeness**
   - Assure that nothing has been omitted from the work-up
4. **Research**
   - Follow a predetermined protocol to meet research objectives
5. **"Education" (curiosity)**
   - Build a personal data base with tests not needed for patient care
6. **Profit**
   - Enhance income by selling testing services or materials (instruments, reagents)
7. **Placebo**
   a. Improve the patient's sense of well-being by obtaining many normal results
   b. Improve the doctor's sense of security with a larger test profile
8. **"Defensive Medicine"**
   - Avoid criticism for failure to order a potentially useful test

Table 2: Traditional Criteria for a Laboratory Test

1. Accuracy
2. Speed
3. Ease of Use
4. Cost

Table 3: Modern Criteria for a Laboratory Test

1. Is the test reliable?
2. Does it provide new information?
3. Will it make any difference?

A candidate procedure must, of course, meet the stringent technical standards expected in an age of high technology. In addition, the test must provide unique information, i.e., patient information not already known from the history, physical examination and prior laboratory studies. A different name and a different range of reference values are insufficient to justify an assay that measures the same physiological function or identifies the same biochemical defect as tests already performed. For example, hemoglobin and hematocrit are expressed in different terms, but both measure the concentration of circulating red cells. If the hemoglobin level is already known, a centrifuged microhematocrit is unlikely to provide additional information. (Automated blood counts performed by modern hematology analyzers provide a full set of red cell indices, but the so-called "hematocrit" is merely a number calculated from the other indices, not an independent measure of the packed cell volume.)

The last and most stringent criterion, the likelihood that knowing the test result will influence the patient's treatment and improve the outcome, would disqualify not only many new assays, but also a sizeable fraction of the well-established tests.

Even clinically useful tests are not always appropriate as in-house procedures. A highly complex analytical method or a requirement for specialized training or equipment may restrict the new procedure to the largest laboratories. Some hospital-based laboratories may be limited by the lack of space to accommodate the new instrument and perform the assay. The test may not be cost-effective unless the costs of setting up the new procedure and continuing support for the equipment, reagents and personnel are balanced by the income anticipated from a realistic estimate of the test volume. Table 4 summarizes the characteristics to consider when deciding whether to offer a new procedure and, if so, whether to perform it in-house or send the specimens to a reference laboratory.

Table 4: Evaluation of a Proposed Test

1. Is it technically satisfactory?
   a. Sensitivity
   b. Specificity
   c. Accuracy
   d. Precision

2. Is it feasible for our laboratory?
   a. Complexity of method
   b. Need for special equipment and/or space
   c. Cost of equipment and supplies, personnel, maintenance, training
   d. Balance of anticipated test volume against fixed costs

3. Is it clinically useful?
   a. Does it provide unique information?
   b. Is the result likely to influence the patient's care or outcome?

Table 5 lists the definitions of some of the important characteristics of a clinical laboratory test. The ultimate clinical utility of a test depends on how it is used in clinical practice. One of the best measures of a test's worth is the *predictive value.* The predictive value expresses quantitatively the probability that the suspected disease is present in a given patient if the result is abnormal or is absent if the result is normal. The calculation is based on Bayes Theorem,[6] and the result can be determined from the test's sensitivity and specificity plus the estimated pretest probability of the disease in that patient. If the patient has no suggestive clinical findings or earlier test results consistent with the disorder, then the default probability is the population prevalence of that disease. Figure 3 illustrates the flaw in many screening programs: Even if the procedure is highly specific, the more uncommon the disease, the more likely it is that an abnormal result will be a false positive.

Table 5: Sensitivity, Specificity and Predictive Value

Sensitivity: The proportion of **positive** test results in patients who have the disease under consideration.

Specificity: The proportion of **negative** test results in healthy subjects or patients who do not have the disease in question.

Predictive Value: Based on Bayes' Theorem. May be calculated if the sensitivity, specificity, and probability of disease (or prevalence) are known.

Positive Predictive Value [PV(+)]:

$$\frac{\text{True Positives}}{\text{True Positives + False Positives}}$$

Negative Predictive Value [PV(-)]:

$$\frac{\text{True Negatives}}{\text{True Negatives + False Negatives}}$$

Fig. 3.

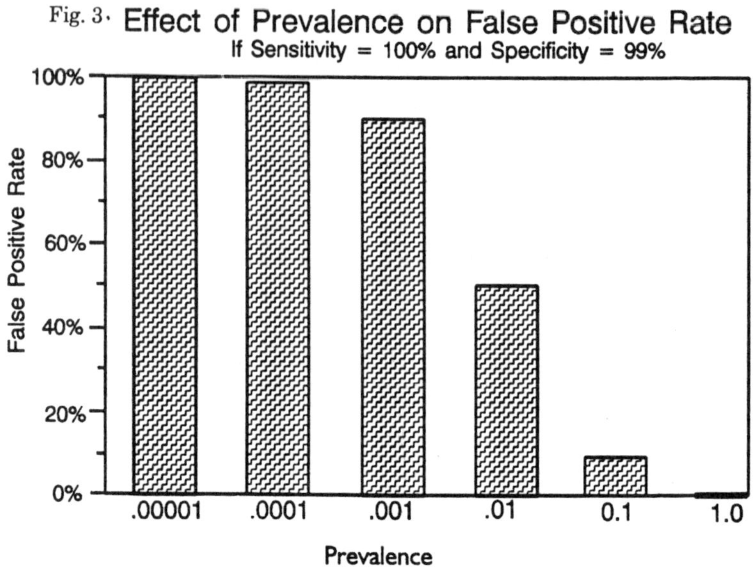

Effect of Prevalence on False Positive Rate
If Sensitivity = 100% and Specificity = 99%
100%
80%
60%
40%
20%
0%
False Positive Rate
.00001
.0001
.001
.01
0.1
1.0
Prevalence

II. BUSINESS PRINCIPLES RELEVANT TO CLINICAL LABORATORIES

The business principles described in the following section will be generally applicable to clinical laboratories of any size (See Biochemistry Section). We have drawn our examples, however, largely from our own experience in the clinical hematology laboratory in an urban university-affiliated county hospital. Our medical center serves a population with a preponderance of socioeconomically deprived residents and recent immigrants. The attending physicians are permanently based at this hospital and hold faculty appointments at the local medical school. Most of the house staff have scheduled rotations through several other affiliated hospitals, including a large tertiary-care medical center, a Veteran's Administration hospital, and a smaller private hospital. The patient populations, case mix and training experience provided at each of these sites vary widely. Local practices and test ordering patterns also show a remarkable range of diversity, even among these closely affiliated institutions.

Our clinical hematology laboratory offers a fairly broad range of routine laboratory procedures (Table 6). The test menu is typical for a general clinical laboratory of average size and encompasses the major categories of hematologic examinations, viz., analysis of circulating blood cells, hemoglobin and hemostasis.

A well-known empirical rule of business is the "80/20 rule", i.e., 80% of the business is based on only 20% of the products. Our laboratory is no exception. Just four high-volume procedures account for more than 80% of our laboratory's in-house workload (Figure 4): (1) Automated complete blood count (with or without an automated differential and platelet count, plus a manual slide review of the peripheral smear for specimens with abnormal CBC results); (2) Urinalysis (including both dipstick analysis and microscopic examination of the urine sediment); (3) Prothrombin time; and (4) Activated partial thromboplastin time. These tests are all performed on fully automated or semiautomated instruments.

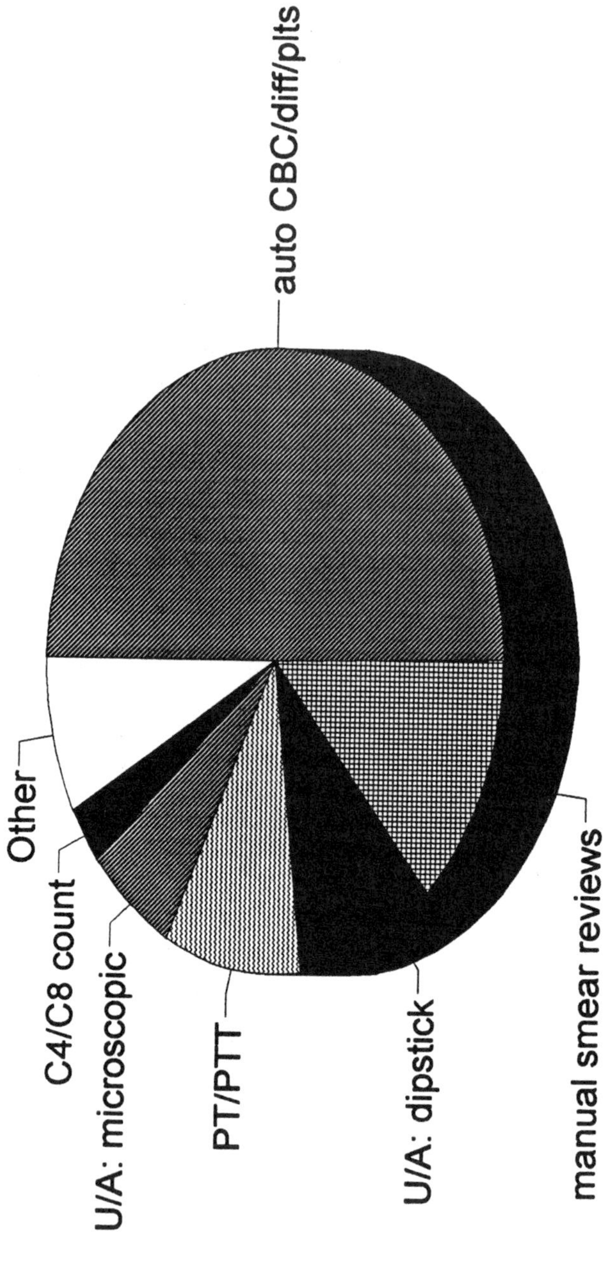

Fig. 4. Hematology Workload at San Francisco General Hospital

Table 6: Tests Performed by the Hematology Laboratory

Evaluation of peripheral blood (automated analyzer)
- Complete Blood Count (CBC)
- Automated differential
- Platelet count

Manual slide review - microscopic examination of a stained peripheral blood smear to evaluate any abnormalities in cellular components as detected on the automated CBC or differential.

Urinalysis -
- Dipstick
- Microscopic - microscopic examination to further evaluated any abnormalities detected by the dipstick analysis

Body fluid analysis - microscopic exam of stained preparation of various body fluids

Erythrocyte sedimentation rate

CD4+/CD8+ T-cell enumeration

Reticulocyte count

Coagulation testing (automated analyzer)
- Prothrombin time (PT)
- Partial thromboplastin time (PTT)

Special coagulation tests - 9 tests

Analysis of hemoglobin - 10 tests

Other specialized hematology tests - 10 tests

The remaining 35 tests performed in our hematology laboratory account for only 12% of the test volume. They are mostly labor-intensive manual procedures that require special training and experience in order to perform the tests and interpret the results. Other tests, which are infrequently requested or require procedures of unusual complexity, are referred to a commercial reference laboratory or to the clinical laboratory of an affiliated hospital (Table 7).

Table 7. Less Common Hematology tests

Coagulation:	Protein C Protein S Antithrombin III Factor V mutation Factor VIII inhibitor titer (Bethesda assay) von Willebrand's panel von Willebrand's multimers anticardiolipin antibodies quantitative Factor II level
RBC structural abnormalities	Sucrose hemolysis Osmotic fragility Ham's test (acidified serum) Ektacytometry
Genetic analysis for hematologic diseases	Chromosome analysis (karyotype) t(15;17) translocation 5q- monosomy 7 inv(16) bcr-abl gene rearrangement T or B-cell gene rearrangement 11q23 gene rearrangement
Leukemia/lymphoma cell phenotyping by flow cytometry	
Other:	Erythropoietin levels Antiplatelet antibodies

Overall cost-effectiveness in the laboratory depends on careful attention to every activity that generates costs or income. The greatest impact, however, will come from the efficiencies achieved with a small number of high-volume procedures.

COSTS OF PERFORMING A TEST WITHIN THE LABORATORY

Many factors contribute to the cost of performing a test in the clinical laboratories. These costs are generally divided into *fixed costs* and *variable costs*, according to their degree of variability in relation to test volume.[7]

Fixed costs include all the expenditures necessary to perform the first test, including the costs of acquiring and maintaining the analytical equipment, the minimum quantity of supplies and reagents, and the minimum number of professional, technical and clerical staff. As the workload increases, the total cost of reagents and consumable supplies will rise, so this expense is classified as a variable cost. Flexible personnel expenses, as for part-time or on- call employees, are also considered as variable costs.

The term s*emivariable or step-variable* is applied to costs that remain constant over a relatively limited workload range, such as the salary of an additional employee who will be hired only after the workload exceeds a specified point, or the cost of an additional analytical instrument.

The *marginal* cost is the incremental cost of performing one more test. Since the personnel, equipment, and other cost components are already in place, the marginal cost is often no more than the cost of the reagents needed for one additional determination.

For highly automated procedures, the marginal cost is generally much lower than the average cost of the procedure, so an increase in test volume may reduce the average cost per test substantially. Conversely, a moderate reduction in the workload will reduce costs only at the marginal rate, and the average cost of the remaining tests will increase. An illustration of the interaction of all these variables is shown in Figure 5.

Fig. 5. Components of Test Costs

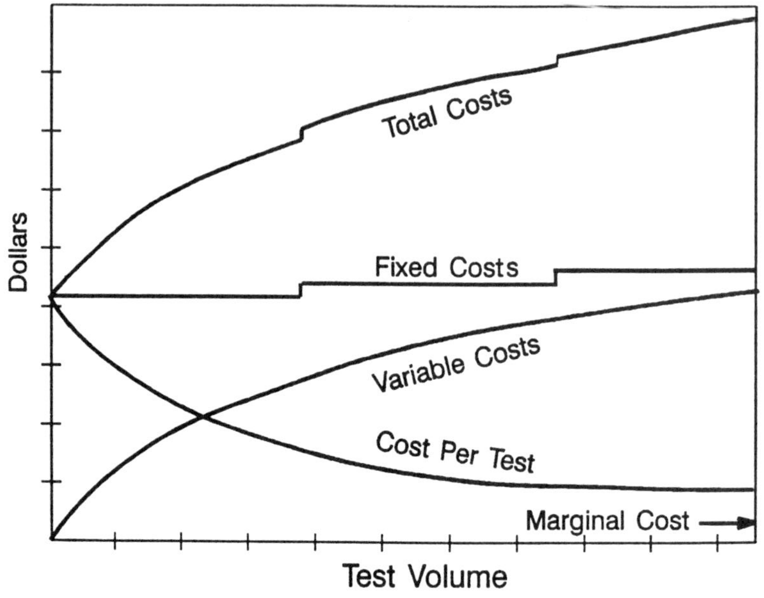

An example of the interrelationship of fixed and variable costs in determining the direct cost of a procedure in the hematology laboratory is shown in Table 8. The equipment needed to perform hemoglobin isoelectric focusing will cost $20,000, and there will be an annual maintenance cost of $2,000. We will assume, for this example, that this equipment is to be used exclusively for this procedure and that it will be paid off over five years (its useful life expectancy) without additional finance charges. The hemoglobin studies may be performed in batches of up to 30 specimens, since each gel plate can accommodate 30 patient specimens plus the controls. If there are more than 30 samples to be studied, the technologist must prepare a new gel plate and run the procedure again. Separate labor and reagent costs can be identified for each sample and for each batch of samples, as shown in the table.

Table 8: Determination of Costs of Hemoglobin Isoelectric Focusing

	Cost Factors	Test #1	Test #30	Test #31
Fixed costs:				
Capital equipment	$20,000.00	$20,000.00	-	-
Annual maintenance	2,000.00	2,000.00	-	-
Variable costs:				
Reagents				
Per run	32.00	32.00	-	32.00
Per specimen	3.00	3.00	3.00	3.00
Labor				
Per run	25.00	25.00	-	25.00
Per specimen	2.00	2.00	2.00	2.00
Totals:		22,062.00	$5.00	$62.00

Average direct cost:	**Specimens per week**	**Average Cost per test***
	5	$39.48
	10	22.38
	20	13.62
	90	8.18

* Based on the assumptions that the capital equipment is amortized over five years (interest charges excluded) and that the laboratory performs at least one batch of tests per week.

The fixed cost of the equipment will be allocated evenly among all the tests performed, so the average cost of each test decreases as the total workload increases. If only one test were performed, the direct cost would be $22,062, but specimens #2 through #30 add only $5.00 each to the total cost. Specimen #31 costs an additional $57.00 because the first gel has been filled and another batch must be run. (In practice, we will probably hold sample #31 until more specimens come in.)

If the laboratory is committed to running the procedure at least once a week, the average cost will range from $39.48 per test (at 5 specimens/week) down to $8.18 (at 90 specimens/week). At a Medicare reimbursement rate of $19.00, it would be less expensive for the small laboratory to send these specimens out to a reference laboratory unless the test volume exceeds the break-even point of 10 - 20 specimens a week.

All of the costs described above are direct costs that can be attributed to the specific activities that generated the expenditure, i.e., one test or group of tests. Expenses for facilities, personnel or equipment necessary to perform the work but not directly traceable to any one procedure are termed indirect costs. Examples of indirect costs include the laboratory computer, library, security personnel, parking facilities, building maintenance costs and rental fees, utilities, and labor costs for administrative and supervisory personnel whose responsibilities cover a broad segment of the laboratory. As it is difficult (by definition) to determine indirect costs, this expense is commonly estimated as a fixed percentage (e.g., 50%) of the direct costs.

Within the clinical hematology laboratory, there are three major components of tests costs: labor, equipment and supplies.

Labor

Labor is the largest cost component of most laboratory tests, so control of labor expenses is the most important factor in running a cost-effective clinical laboratory. The greatest cost savings will occur with a decrease in the personnel time needed to perform a high-volume test.

To evaluate and reduce labor costs, four major issues must be considered: 1) the test menu, 2) the number of work sites and the laboratory layout, 3) selection of appropriate personnel for each task, and 4) orientation, training and retention of employees.

Test Menu: The number of tests offered by the laboratory and their complexity will largely determine the staffing requirements. A test menu offering a small number of predominantly automated tests may require only a few technical personnel, whereas a large variety of tests performed by labor-intensive manual procedures will require a larger number of technical personnel with more extensive training and experience.

Number of Work Sites and Layout: Satellite laboratories are often set up to serve remote locations or to perform special functions, e.g., "stat" testing, outside of the main laboratory area. The need for such smaller laboratories physically separated from the main laboratory is site-dependent and often highly controversial.

On the one hand, turnaround time can be dramatically reduced by a strategically located satellite laboratory, particularly in medical centers with widely scattered patient-care units and poor systems for transporting specimens. The main impediment to rapid turnaround time is generally not the analysis itself, but the time required to deliver the specimen to the laboratory and the report to the physician.[8]

On the other hand, the smaller the number of work sites, the fewer supervisors and technical personnel are required to do the work and the easier it is (at least in theory) to manage the personnel. Moreover, satellite laboratories usually require some duplication of equipment as well as more personnel to cover all hours of operation at all laboratory sites.

Compared with the main laboratory, the satellite laboratory is more likely to be staffed by a small number of technologists and have wider fluctuations in the workload, leading to swings between an overwhelmed staff producing reports with unacceptably long turnaround times and nonproductive personnel waiting for the next test request.

Our experience has shown that whenever possible, it is more effective to retain as many activities as possible within the main laboratory. One useful approach is to designate a specific area for "stat" test requests and then incorporate these "stat" specimens into the workflow alongside the routine specimens. Personnel normally assigned to routine hematology procedures (e.g., hemoglobin electrophoresis) can then be reassigned to temporary duty on the "stat" bench during peak periods.

Consideration of workflow in the design of the hematology laboratory will have a great influence on overall efficiency. The automated blood count accounts for the largest proportion of tests requested in most hematology laboratories and is the procedure most often requested stat. The specimens can be handled and processed more efficiently if the automated CBC analyzer is located adjacent to the specimen-processing area or near the door through which the specimens arrive.

In contrast, tests ordered less frequently or performed only during the day shift can be performed at bench locations farther from the laboratory entrance. The laboratory layout should also accommodate the special needs of the evening and night shifts in the 24-hour laboratory. The smaller crew generally assigned to these shifts will be able to work more effectively when all the equipment needed for the tests performed on these shifts is in close proximity.

Selection of Personnel: Labor costs vary not only with national and local market conditions of supply and demand, but also according to educational requirements imposed by local, state or federal standards. Higher educational requirements generally require more highly compensated personnel. The Clinical Laboratory Improvement Amendments of 1988 (CLIA '88) require virtually every facility performing tests on human subjects to have a Federal laboratory license. The regulations also specify the minimum requirements for education and experience for personnel performing laboratory tests at each level of complexity.[9] Individual states or large cities may impose even more stringent requirements.

Awareness of the federal and local laws regulating technical personnel and use of the least expensive labor for a particular job will maximize savings on labor costs while avoiding inadvertent violations of the law.

In California, for example, unlicensed technical personnel (often known as "technicians" in other areas) are not permitted to perform complex tests. They are allowed, however, to operate instruments that have a mechanical recording mechanism (e.g., automated CBC analyzers and automated urine dipstick readers), provided that a licensed technologist reviews and files the results.

The difference in monthly salary between unlicensed and licensed technical personnel at San Francisco Bay Area hospitals ranges from $700 to $1700 a month. Thus, assigning the permitted procedures to unlicensed personnel instead of licensed technologists can yield a significant reduction in labor costs.

The most efficient use of labor requires that all personnel have enough work to do at all times. On the other hand, the workforce must be large enough to cope with peak workloads and emergencies without developing a backlog with unacceptably long turnaround times. A good way to balance these needs is to separate those tasks that must be performed immediately from those that can be done later.

The entire technical staff, including supervisors when necessary, can be enlisted to handle the benchwork during the peak hours (typically 1 - 2 hours each day). After the urgent work has been completed, the staff can return to their previously assigned routine tasks.

During periods of lower activity, such as during the evening or night shift, it is reasonable to schedule major instrument maintenance, such as retubing a cell counter. Since there is always a possibility of an unforeseen instrument problem, however, appropriate back-up equipment must be available or the maintenance should be performed by a highly skilled factory-trained person.

In our hospital, the hematology section of the laboratory is administratively responsible for hospital-wide phlebotomy service. This arrangement offers a special advantage that may not be immediately obvious, viz., the laboratory can predict its short-range staffing requirements from the number of specimens scheduled for collection by the phlebotomy service.

The laboratory must be alert to changing workload patterns and maintain the necessary flexibility to adjust to changing needs. For example, our laboratory has been providing pediatric hematology services through a small satellite laboratory adjacent to the pediatrics clinic.

We observed a steady decline in the workload at that site over the past few years, until the technologist's time was no longer well utilized. We therefore gave the technologist a pager, and this employee now works in the main hematology laboratory until the pager indicates that his/her services are required in the satellite laboratory. A similar plan may be practical for any off-site laboratory at which the technologist or phlebotomist is underutilized.

Adjustments in staffing in response to peaks and valleys in workload volume may require moving personnel from one shift to another or changing shift times. Personnel utilization on all shifts can best be directly assessed by the supervisor who spends some time on all shifts and performs benchwork to obtain first-hand information. Alternatively, the necessary information can be gathered from the personnel permanently assigned to each shift. Suggestions volunteered by the staff in the spirit of a "team approach" may have the best chance of acceptance and implementation.

Orientation, Training and Retention of Employees: Orientation is an opportunity to introduce the individual to the organization and welcome the new employee as a member of the team. The employee should be aware of the expectations of the laboratory, including work hours, biosafety regulations, and, most importantly, attendance. Orientation of phlebotomists is especially important, since they are often the only laboratory personnel the patient ever sees and they must fully understand their role as the laboratory's "ambassadors."

During the training period, the employee should be given a list of the tasks to be mastered. Formal training should be documented with a log that includes the dates of training, satisfactory completion of the training tasks and the trainer's name. The employee should not be permitted to perform procedures for patient care until the formal training has been completed and the trainee feels competent.

At least 4 - 6 weeks is needed to train a new employee adequately, and trainees are not fully productive for several more months. Training costs are expensive, so it is essential to minimize employee turnover. Problem employees should be eliminated as early as possible during the probationary period in order to avoid greater expense later.

If the performance of a long-term employee changes suddenly, personal problems should be considered. The employee should be counseled confidentially and referred to the appropriate counseling service as soon as possible.

Equipment

The widespread use of automated instruments in the clinical laboratory has dramatically reduced the labor required to perform tests in most sections of the clinical laboratory. For hematology, automated instruments are available for performing complete blood counts, coagulation studies and urinalyses. Automation has increased the laboratory's capacity to accommodate growing workloads and has reduced the average cost of a laboratory test. The obvious trade-off is the high initial expense and continuing maintenance costs of these complex instruments.

Three desirable features for any automated analyzer are automated sampling, barcode reading capability and a direct interface with the laboratory computer system. Each of these features helps to avoid human error and reduces labor costs. Another desirable feature is the ability to select one or more tests from instruments capable of performing multiple tests simultaneously. For example, a clinician monitoring the effects of a blood transfusion may be interested only in the post-transfusion hemoglobin value and none of the other results routinely reported as part of the automated blood count. Reports of irrelevant (and worse, unrequested) tests may divert attention from more important results and create a real problem if the result is unexpectedly (and perhaps falsely) abnormal.

As described in the clinical chemistry section of this book, a laboratory can acquire automated analyzers and similar instruments in a number of ways. The most obvious means is outright purchase. A second is the lease or lease-purchase, which permits spreading the cost of the instrument over a defined period of time (generally 3 - 5 years or the estimated useful life of the instrument). A third kind of arrangement is the "reagent rental," in which the instrument is provided by the reagent vendor at no additional cost, in exchange for an exclusive contract for the reagents used with that instrument. Nothing is free, of course, and the cost of the instrument will be hidden in the price of the reagents. The negotiated rate is typically based on the anticipated test volume, with a second (lower) rate that will be honored if the volume is higher than expected.

An important principle to be considered in the purchase decision is the "time value" of money, viz., a dollar spent tomorrow is not as expensive as a dollar you have to spend today. A variety of formulas exist to help an individual laboratory determine its most cost-effective option.[10] No matter which arrangement is selected, however, the final cost will depend to a considerable degree on the effectiveness of the negotiator.

It must be emphasized that new technology is not necessarily better, and the results are not necessarily clinically useful. For example, a recent generation of whole blood coagulation analyzers can report a calculated fibrinogen level along with the prothrombin time. These derived values correlate well with those obtained by the reference Clauss method only within the normal range; outside this range, the correlation is poor.[11,12] Instruments of this type are attractive primarily because of the rapid turnaround time. They may find a use as a screening device, separating the normal specimens from the small proportion that require further testing by a slower but more reliable procedure.

Supplies and Reagents

When the automated analyzer is acquired by any means other than reagent rental, the laboratory will have to decide whether to purchase the reagents from the instrument manufacturer or from another vendor. The manufacturer's reagents are often more expensive, but they have been tailored to the specific requirements of the instrument and are certified to produce accurate results and not to damage the instrument. Products from other sources, while usually less expensive, may or may not be identical to the instrument maker's reagents.

In making its purchasing decision, the laboratory will have to weigh the potential cost savings against the need to spend more time evaluating the accuracy and safety of the alternative products, as well as the need to consider one more variable when troubleshooting a malfunctioning instrument. (See clinical chemistry section for additional discussion of this topic)

Many of the materials needed for tests performed manually in the clinical hematology laboratory, e.g., occult blood in stool, urine pregnancy tests and peroxidase stains, are now available in kit form. The test kits usually contain just enough materials to perform a single test plus positive and negative controls. Although the price of kits is relatively high compared with that of bulk reagents, they are generally fast and convenient. Moreover, the higher cost may be offset by the savings in labor and perishable supplies that would otherwise be required for these low-volume tests.

III. OTHER COST CONSIDERATIONS

"STAT" TESTS: ARE THEY WORTH THE EXTRA COST?

The cost of running a "stat" test is higher than that of a routine test -- sometimes far higher -- because each procedure is performed individually and the benefits of batch testing are lost. The added costs for labor (including those involved in transporting the specimen and reporting the results) and supplies (including extra controls and standards) will depend on the nature of the procedure and the potential batch size for routine testing. Some of these costs will be difficult to measure directly, including those associated with the delayed reporting of other patients' results, whose tests have been given lower priority.

An example of the extreme differences in costs that may be encountered between routine and "stat" testing is shown in Figure 6. This example is from the chemistry laboratory, a combination of six tests known locally as the "liver panel." Routine testing is performed in large batches on a highly automated, computer-based biochemical analyzer (Miles DAX), whereas "stat" tests are done on another instrument that has higher reagent costs (Kodak Ektachem). Most of the added cost of the "stat" procedure, however, is associated with the increased labor expense of handling the specimen and the results individually.

It should be borne in mind that, regardless of the cost to the laboratory, the published prices of laboratory tests vary considerably. The official list price may bear little relationship to the discounted fee paid actually by the patient or third-party payer. For the test combination shown in Figure 6, the list price may be as high as $100 or even $150.

Fig. 6. Prices and Costs of a Liver Biochemistry Test Panel

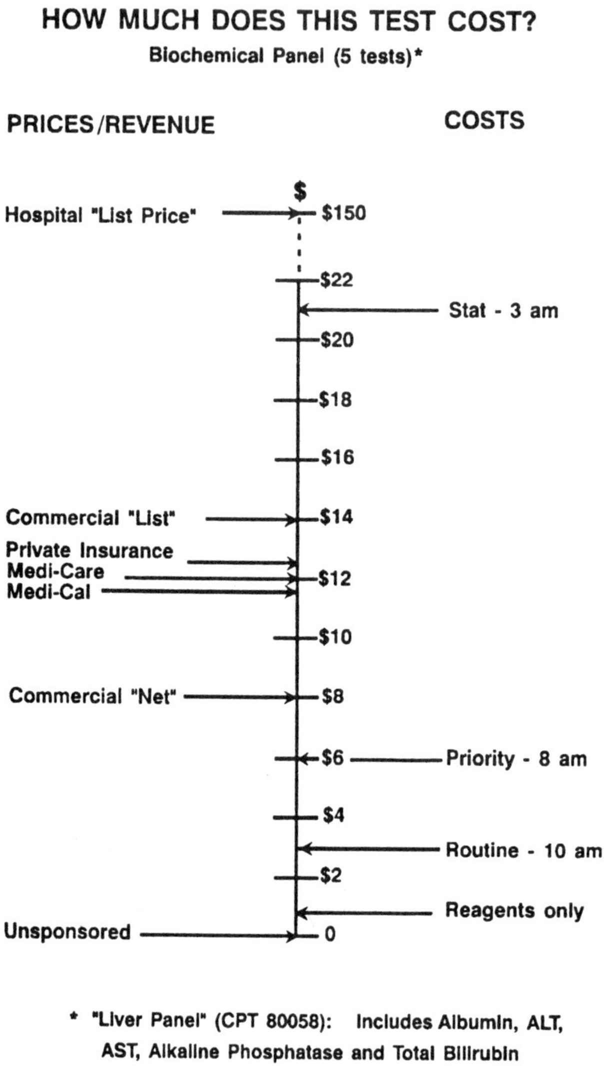

The tests most often ordered "stat" in the hematology laboratory, in contrast to those in the chemistry laboratory, are performed on automated analyzers that can accommodate individual samples or batches equally well. For this reason, differences in analytical cost between routine and "stat" tests may be much smaller, especially if the results are normal. For example, in order to run a "stat" automated blood count, the technologist interrupts the current run to replace the next specimen in line with the "stat" specimen, manually enters the new specimen identification information, then retrieves and reviews the results.

If the results are flagged as abnormal, the technologist must stop to prepare, stain and review a peripheral blood smear in order to verify the report.

Running the extra sample on the automated instrument may take an extra minute of the technologist's time. Making, staining and reviewing an abnormal peripheral smear may take additional 5 - 10 minutes of time and interrupt the normal flow of work in the laboratory. Thus, most of the extra cost for "stat" testing is associated with the work necessary to verify abnormal results.

It seems generally true in the hematology laboratory that the difference in cost between normal and abnormal specimens is greater than that between routine and "stat" testing. The average test cost will be strongly influenced by the case mix of patients served by the laboratory. Laboratories in medical centers serving more severely ill patients will encounter more abnormal results, which in turn will consume more resources in verifying the findings.

In contrast to the direct analytical expenses, it is more difficult to measure some of the other costs associated with "stat" testing, especially those incurred outside the laboratory. If the "stat" procedures cause a delay in reporting the results of all the specimens processed routinely, this may lead to subsequent delay for another patient in the start of appropriate therapy, discharge or transfer to a skilled nursing facility.

Conversely, some "stat" tests, despite the increased laboratory expense, may provide life-saving benefits for the patient and overall savings for the hospital when critical case management decisions depend on the laboratory results.

For example, quantitative fibrinogen levels are determined by a labor-intensive manual method. When a "stat" fibrinogen level is requested, a technologist assigned to routine procedures must put aside the routine work in order to perform this test. For the patient, the result may mean (if the fibrinogen level is low and the patient is bleeding) timely administration of appropriate blood products, avoidance of a severe hemorrhagic episode and consequent complications; or (if the fibrinogen level is normal) avoidance of inappropriate blood product transfusion and prevention of unnecessary exposure to transfusion-transmissible infectious agents. Either result may save the hospital money by decreasing the length of hospitalization, administering fewer costly blood products, and avoiding the need for other diagnostic and therapeutic services.

A similar example is the rapid determination of specific coagulation factor levels to help guide therapy in a patient with hemophilia who is about to undergo a surgical procedure. Even though the laboratory procedures for specific coagulation factor levels are labor intensive and therefore costly when performed "stat," the potential benefits are very high, as appropriate treatment with the relevant factor may prevent surgical complications, assure an uncomplicated recovery, and shorten the patient's hospital stay.

ALTERNATIVES TO "STAT" TESTING

The goal of "stat" testing is to have test results available as soon as possible. In many laboratories, the target turnaround time (TAT) is about 45 minutes, from receipt of the specimen in the laboratory to reporting the results. (Additional time -- often extensive -- is needed to transport the specimen to the laboratory or deliver the report to the ordering physician.) It has become customary in some areas for physicians to order tests "stat" when immediate results are not needed to resolve a life-threatening condition or even to decide on a course of treatment. In some sections of our laboratory, as many as 30 to 50% of all test requests are "stat" orders. Therefore, alternatives to "stat" testing were considered.

Priority Testing

When we studied the "stat" ordering patterns at our hospital, we found that in some instances the routine reports were arriving too late for ward rounds, or that the results were needed soon but not immediately, e.g., within 2 hours. In order to meet these needs while reducing the unnecessary costs associated with "stat" testing, we instituted an intermediate "priority" level of service, with a target TAT of 2 hours (Table 9).

Table 9: Test Priorities*

Routine:	$\leq$ 8 hours
Priority:	$\leq$ 2 hours
Stat:	$\leq$ 45 minutes
Point-of-Care:	$\leq$ 10 minutes at bedside $\leq$ 1 hour if result is beyond linear range

* Target turnaround time for laboratory procedures at San Francisco General Hospital

We also made individual arrangements for levels of service tailored to meet the particular needs of individual wards and clinics accustomed to ordering a large number of "stat" tests.

For example, all "priority" specimens received from the intensive care unit by 8:00 am will have results reported by 10:00 am, in time for the physicians' ward rounds. This approach has been remarkably effective in reducing unnecessary "stats" and enabling the laboratory to improve its work flow.

It is important to monitor the turnaround time for each category of test periodically in order to verify that the laboratory is meeting its stated goals. The consequence of failure to report routine and "priority" test results in a timely manner is generally an unwelcome increase in the number of "stat" requests.

Point-of-Care Testing*

Testing carried out at or near the bedside or clinic examining room, often with small portable instruments designed specifically for that purpose, is becoming increasingly popular. The concept is in reality a return to an earlier era of laboratory testing, before large automated instruments in central laboratories became dominant. The appeal of point-of-care testing is based on its presumed convenience, moderate cost and improved turnaround time compared with tests that require the specimen to be sent to the main laboratory.

The most common example of point-of-care testing in our hospital is the periodic measurement of blood glucose by fingerstick in patients with diabetes mellitus, just before the next scheduled dose of insulin. The results are available within a few minutes, compared with an hour (for a stat test) to several hours (for a routine test) for a venous blood specimen sent to the main laboratory.

* For another discussion of Point-of-care testing, see Chemistry Section.

Point-of-care testing has added another layer of complexity to the work of the laboratory. Unless otherwise specified (and separately licensed), any laboratory testing performed in the hospital setting is presumed to be the responsibility of the director of the main laboratory.

The CLIA '88 regulations classify glucose as a waivered test and apply the least rigorous rules for education and training of the personnel performing the test. The more stringent CLIA '88 rules applicable to laboratories of higher complexity may pertain to other point-of-care procedures such as blood gas analysis or prothrombin time.

Regardless of the CLIA '88 rules, however, any test performed in a hospital setting must also meet the requirements established by the Joint Commission on Accreditation of Healthcare Organizations (JCAHO). Additional regulations, including those of the state licensing authority, also apply.

This web of regulations covers all aspects of point-of-care testing, including training and certification of the personnel, calibration and maintenance of the instruments used to perform the test, controls, standards, full documentation, and periodic proficiency testing for each instrument used in the program.

Until a few years ago, our nursing staff performed fingerstick glucose determinations with little regard for formal regulations. In the face of continuing clinical demand for bedside glucose testing, the new CLIA '88 rules created a serious problem for a nursing staff reluctant to deal with unfamiliar and often complex regulations.

At this point, the Clinical Laboratories were asked to help develop a new program in collaboration with the nursing department. The laboratory assumed responsibility for its traditional areas of expertise, i.e., interpreting the regulations, administering the quality assurance program, and calibrating and maintaining the portable glucose meters. The nursing department retained responsibility for training and periodically recertifying the nurses designated to perform the test.

When we reviewed the program soon after its inception, it was apparent that the expense of training and recertifying many hundreds of nurses would be substantial. A substitute plan was developed that would not only cost less, but also make much better use of the available resources.

Instead of training the nurses in every area of the hospital to perform fingerstick glucose testing, we identified a few locations that had higher needs than the rest, e.g., intensive care units and diabetes clinics. Under the new plan, nurses perform the procedure only in these high-volume sites. Patients in other areas of the hospital are tested by an expanded phlebotomy team in the course of their regularly scheduled rounds. The additional training and supervision of the phlebotomy team remain the responsibility of the laboratory.

The money saved from the reduced scope of the nursing department's glucose testing program not only paid for the cost of bedside glucose testing by our phlebotomists, but also allowed us to expand routine phlebotomy service from two to four scheduled pickups a day, timed to coincide with requests for glucose testing before meals and in the evening.

Another advantage of the laboratory's direct participation in the glucose testing program is improved follow-up for out-of-range results. When a test result falls within the relatively narrow linear range of the portable meter, the value is recorded in the patient's chart. If the results are outside of this range, the phlebotomist can immediately collect a venous blood sample and deliver it to the main laboratory for a glucose determination by the standard method. The definitive value is reported within an hour.

The collaborative program has enabled our hospital to continue providing a pre-existing service with improved quality, full compliance with governmental regulations, and the added benefit of increased regular phlebotomy service. If bedside glucose testing had simply been discontinued instead, we would probably have experienced a massive increase in requests for "stat" glucose.

The success of the bedside glucose plan has encouraged us to address formerly independent point-of-care testing programs in other parts of the hospital. The surgery department, for example, has been using instruments for blood gas analysis and rapid prothrombin time determination in a space adjacent to the operating room. These analyzers are now calibrated and maintained by clinical laboratory staff, and the surgical personnel who perform the tests participate regularly in quality assurance programs administered by the main clinical laboratory.

The recent growth of point-of-care testing has been encouraged by the manufacturers and vendors of portable instruments. They tend to emphasize the speed and convenience of the procedure while minimizing the costs and problems of calibration, controls, instrument maintenance, and documentation of the results.

Moreover, many of the portable analyzers marketed for bedside testing have a short range of linearity and are suitable only for screening. Abnormal results must be confirmed by a standard method.

The potential purchaser may be impressed by the low initial instrument cost and fail to notice the high cost of the disposable supplies (e.g., special cartridges for prothrombin time testing) or the time that patient-care personnel will have to divert from other duties in order to test the patient specimen (plus controls) and record the results.

When a bedside test is compared directly with standard testing in the main laboratory, the added clinical benefits (primarily faster turnaround time) do not always justify the increased cost.

Participation by the clinical laboratory in the hospital's point-of-care testing programs is strongly recommended as a means of assuring consistency in programs and procedures, adherence to appropriate quality assurance programs, and timely confirmation of abnormal values by the main laboratory.

The laboratory's expertise may also contribute to direct monetary savings, by avoiding unnecessary duplication in training programs or services and by discouraging unnecessary expenditure for inefficient or ineffective bedside tests.

IV. NEW OR UNCOMMON TESTS

DIAGNOSTIC TESTS BASED ON MOLECULAR BIOLOGY PRINCIPLES

New laboratory procedures are continually being developed. It is the responsibility of the laboratory director to decide which of these tests are clinically useful and should be offered by the laboratory. This is best done in collaboration with interested clinician-users, either informally or following a formal protocol established by the institution (e.g., the hospital's laboratory advisory committee).

From time to time, a new test emerges that is clearly superior to the methods previously available for the same clinical indications. Sometimes the improvement is based on the increased precision of an automated instrument over that of the corresponding manual test, as in automated blood cell counting. In other cases, the new procedure is based on entirely new principles with intrinsically higher sensitivity and/or specificity than the current test. In either instance, the laboratory's limited resources should be conserved by discarding the obsolete method as soon as possible instead of merely supplementing it with the new one.

The laboratory director should be prepared to encounter resistance from clinicians who become surprisingly attached to the old familiar tests, regardless of their demonstrated inferiority.

Recently, there has been a surge of interest in tests based on the principles of molecular biology for the diagnosis of hematologic disorders (Table 7). The methods are labor intensive and require expensive reagents. Aside from the high cost of this group of tests, their major drawback is the lack of agreement on the clinical interpretation of the results.

These tests determine genotypic abnormalities, whereas the diagnosis of disease has traditionally been based on the observation of morphologic (i.e., phenotypic) abnormalities .

As would be expected, the genotypic abnormalities identified by the new molecular methods do not correlate perfectly with the observed phenotypes, creating a diagnostic dilemma in the atypical cases. This problem is most apparent in diseases that are closely associated with a specific molecular abnormality.

As an example, the Philadelphia (Ph^1) chromosome, a translocation between chromosomes 9 and 22 [t(9;22)], is a karyotypic abnormality observed in 90% of cases of chronic myelogenous leukemia (CML). The molecular lesion is the translocation of the "abl" gene into a different chromosomal site, the "bcr" locus, with subsequent dysregulation of the expression of an altered "abl" gene product.

Transgenic mice expressing the bcr-abl fusion product develop a disease phenotypically identical to CML, but it is not known whether the altered "abl" gene product itself is responsible for the pathogenesis of CML.[13]

Now a new test for the bcr-abl gene rearrangement is being marketed. This test is slightly more sensitive than the standard Ph^1 karyotype, as 95% of patients with a clinical diagnosis of CML have a demonstrable bcr-abl gene rearrangement.[14]

The remaining cases, though morphologically and clinically identical to CML, lack the bcr-abl gene rearrangement. It is unclear how they should be classified. It has been long recognized that the subset of CML patients who lack the Ph^1 chromosome (and now, the bcr-abl gene rearrangement) have a poorer prognosis than the typical patient. They may in fact have a different disease, thereby explaining the poor response to conventional CML therapy.

As another example, population studies have defined specific karyotypic abnormalities associated with a better prognosis [e.g., 5q- in myelodysplasia, (inv^{16} in myelomonocytic leukemia with eosinophilia] or a poorer prognosis (e.g., multiple karyotypic abnormalities in myelodysplasia, 11q23 translocation in lymphoid or myeloid leukemias).

In contrast to the general trends demonstrated in population studies, the ability of any laboratory procedure to determine a specific patient's prognosis will be limited by individual variability. Moreover, testing for this purpose is clinically useful only if the result will help the physician select a more appropriate mode of treatment.

Further progress in diagnostic tests based on molecular biology may lead to profound changes in the approach to medical diagnosis. The methods are still evolving, and at present their interpretation remains controversial.

The procedures described here are complex and labor intensive. They require the skills of highly trained personnel and the resources of a well-equipped research or reference laboratory.

The average cost is about $200, 10 to 100 times higher than that of a routine hematology test. Further study is needed before this expensive group of tests can be considered ready for clinical application in the diagnosis or management of disease.

USE OF REFERENCE LABORATORIES*

When a new test is approved, it will also be necessary to predict whether the anticipated test volume justifies the cost of setting up the procedure in the laboratory. If not, these test specimens should be sent to a reference laboratory.

The cost-conscious laboratory should periodically reevaluate its menu of tests and consider whether they should all remain "in-house" procedures or whether some are suitable candidates for referral to an outside laboratory. The decision to send out selected specimens depends on a number of factors other than cost alone, including the turnaround time required for optimal patient care and the stability of the specimens during the interval between collection and delivery to the reference laboratory.

As an example, the typical clinical hematology laboratory receives a large number of blood specimens for a complete automated blood count, including an automated differential and platelet count. The analytes in a specimen collected in an evacuated tube containing EDTA anticoagulant are stable for 24 hours at room temperature.

If the reference laboratory is located in a nearby place, so that routine turnaround time from the reference laboratory is no longer than 18-24 hours, it will be adequate to meet the needs of most clinic patients and some inpatients in stable condition.

On the other hand, many hospitalized patients may have clinical conditions subject to sudden deterioration, e.g., active bleeding, and immediate test results are required in order to assure appropriate treatment. For this reason, an on-site automated hematology analyzer capable of providing fast and accurate results is indispensible for any medical facility that treats acutely ill patients.

* For a discussion of the use of reference laboratories from a clinical chemistry point of view, see Chapter I.

Once the laboratory has acquired an instrument with adequate capacity to handle routine specimens as well as the "stat" requests, the additional expense is limited to the low marginal cost of reagents and labor. The higher the overall test volume, the lower the average cost per test. Thus, it is highly cost-effective to offer automated blood counts as a routine "in- house" procedure.

In contrast, many of the diagnostic tests employed in the workup of patients with an unusual hemoglobin variant or coagulation disorder are slow, complex, and labor intensive. They may require specialized training and/or equipment, and interpretation of the results may require a level of training and experience available only in large laboratories. For some or all of these tests (depending on the anticipated test volume, the available in-house resources and special interests of the medical staff), it will probably be more cost-effective to use a reference laboratory than to attempt to set up the procedures on site.

The argument has sometimes been used, especially in teaching hospitals, that an excellent laboratory must be able to perform every variety of test in-house, and that the institution's prestige depends upon the availability of esoteric procedures. In our view, this concept has no place in an environment that demands proof of clinical utility and optimal cost-effectiveness.

It is important to remember that the medical director of the laboratory is responsible for the quality of all test results, even if the testing is performed at an outside laboratory. The selected reference laboratory should be asked to provide full information about the methodology used, the reference range (and the basis of its derivation), and its records of proficiency test performance and compliance with federal and local regulations.

Critically important factors to consider in the selection of a reference laboratory include the turnaround time for routine results and the availability of "stat" testing when necessary; whether an on-line computer terminal is available for rapid results reporting; hours of availability of courier services; and the total charge for routine and "stat" tests, including any hidden "extras."

The in-house laboratory personnel should be able to reach designated technical personnel at the reference laboratory easily in case questions arise regarding the required specimen, sample stability, test protocols, quality assurance issues or interpretation of results.

There is reason for concern when the reference laboratory is chosen primarily on the basis of competitive bidding, or when a new and unfamiliar company wins the contract. It is prudent to learn as much as possible about the operations and general reputation of an unknown laboratory.

An extremely useful technique for choosing a reference laboratory is to split a number of specimens with known test values and send the aliquots to several laboratories for comparison of the accuracy of results, turnaround time, and availability and courtesy of the staff. NCCLS has developed a checklist of concerns for the clinical laboratory faced with the selection of a reference laboratory.[15]

In addition to the more common tests, reference laboratories often offer a wide range of more unusual procedures. As explained earlier, many of these tests have very limited clinical utility and should be ordered only under special circumstances.

The following are examples of tests that are useful in specific clinical indications but which are often ordered inappropriately: (a) Erythropoietin level, useful primarily in the evaluation of a patient with HIV-1 infection and anemia, who may be a candidate for exogenous erythropoietin therapy; (b) Protein C, Protein S and Antithrombin III levels, useful mainly for evaluation of the family of a patient with a suspected hereditary hypercoagulable state.

In order to avoid overuse of tests with limited clinical utility, we have developed a protocol under which requests for unusual procedures must be reviewed and approved by a laboratory-based physician in consultation with the patient's physician and/or the clinical hematology service. This process weeds out requests for tests when the results will be unable to help make a diagnosis or influence case management.

For the unusual case of a disagreement over the clinical utility of a new test (or over a laboratory policy) that cannot be resolved between the laboratory director and the clinician, an appeal process should be available. At our hospital, this function is filled by the laboratory advisory committee, a representative group of clinical and laboratory-based physicians appointed by the chief of staff. When necessary, this committee can serve as an impartial jury to hear each side, weigh the evidence, and provide a binding decision.

V. SUMMARY AND CONCLUSIONS

COST-EFFECTIVE STRATEGIES FOR THE CLINICAL HEMATOLOGY LABORATORY

Under the traditional fee-for-service practice of medicine, it is customary to bill patients for each laboratory test. As in other areas of medical practice, clinical laboratories have long been accustomed to cost-shifting, i.e., setting charges higher than the cost of performing the tests for some patients, in order to cover losses in other areas such as those incurred when providing care to indigent patients whose medical costs are inadequately reimbursed.

In general, the standard charges for tests done by hospital laboratories have been high enough to cover not only the laboratory's own operating costs, but also those of cost centers elsewhere in the hospital.

In the past, the laboratory's profitability was sufficient to place the laboratory in a favorable position in the annual competition for funding. This is no longer true.

Medical practice is undergoing profound changes, with increasing emphasis on managed care and shrinking budgets. More than ever, administrators demand evidence that the laboratory is operating economically in comparison with similar laboratories. Some hospital laboratories may be faced with the threat of an obvious alternative, viz., simply closing the laboratory and sending all of its lab work to an outside laboratory.

Before a plan is formulated to reduce costs, it is important to know not only the overall cost of running the laboratory, but also each of the fixed and variable components, as described earlier. If a test is seldom ordered "stat," the volume is low, and/or the marginal cost is not substantially less than the reference laboratory's charge for the same procedure, it may be cost-effective to send these specimens to the outside laboratory.

Effective use of personnel is one of the most powerful means of reducing costs. The laboratory should always take advantage of advances in technology that eliminate or simplify labor-intensive procedures. Non-specialist personnel should be used for routine tasks instead of more highly trained and highly paid technologists whenever it is technically (and legally) feasible.

In our hematology laboratory, laboratory assistants now perform a number of tasks formerly done by technologists in connection with the operation of the high-volume hematology analyzers.

We also replaced manual differential counts with automated differentials, adding a qualitative slide review (but not a manual count) when necessary to verify abnormal results on the automated blood count.

We now do reticulocyte counts on the automated hematology analyzers that are already available in the laboratory and familiar to our personnel. Compared with the standard manual reticulocyte method, the automated procedure offers vastly improved precision along with the labor savings. It also provides a more rapid turnaround time than the corresponding method based on a general-purpose flow cytometer.

Changes like these have produced substantial savings in our laboratory with no compromise in quality. Indeed, the automated differentials alone saved the labor equivalent of two full-time technologists.

Perhaps the most effective (but admittedly the most difficult) way to reduce laboratory costs is to educate the users, i.e., the physicians who order the laboratory tests, so that they avoid ordering unnecessary tests and refrain from ordering "stat" tests unless there is a genuine clinical need for rapid turnaround time.

A physician who knows how to make the best use of the laboratory's resources to make a diagnosis or manage a patient can make an important contribution to the laboratory's efforts to hold down costs.

In our laboratory, for example, microscopic examination of the urine sediment is performed on half of the specimens submitted for urinalysis. The microscopic examination, however, is unlikely to yield any useful information unless the dipstick produces a positive result or there are clinical signs suggesting urinary tract disease. Understanding of this principle can dramatically reduce the number of microscopic examinations ordered solely for screening purposes.

Improving physicians' awareness of the high cost of "stat" tests is one way to encourage acceptance of the principle that "stat" tests should be ordered only when the patient's condition warrants the extra effort and the results are likely to influence the patient's outcome.

Practice guidelines developed with an understanding of the principles relevant to laboratory testing can lead to more cost-effective laboratory utilization. As an example, the efficacy of a platelet transfusion can be estimated by means of a platelet count taken one hour after the transfusion, but not 24 hours later.

Similarly, the dosage of coumarin for anticoagulant therapy is adjusted according to the results of serial prothrombin time measurements, but interpretation of the test results requires familiarity with the pharmacokinetics of coumarin (viz., a half life of 31-58 hours and a steady state level attained only after 4-5 half-lives).

There is a great need for computer-assisted order entry systems that can provide information about testing algorithms, assist in the selection of appropriate tests and block duplicate orders.

Procedures that permit laboratories to compare their costs to those of other laboratories of similar size and patient mix may be very helpful both in self-assessment and in interactions with hospital administrators.[16] Comparative cost analyses may also facilitate the identification of specific problem areas in the laboratory.

Now, more than ever, clinical laboratories are being challenged by the opposing pressures of declining budgets and increasing costs. The expansion of managed care and competitive bidding by providers is almost certain to have an adverse effect on the funding of clinical laboratories. On the other hand, the implementation of new laboratory techniques (e.g., molecular diagnostics) and compliance with new regulations (e.g., CLIA '88) are generally associated with increased costs.

One path to survival in the face of such pressures is for the clinical laboratory staff to play a more prominent role in the hospital or medical community within their area of special expertise. One such area is the education of physicians in appropriate and cost-effective utilization of the limited resources available for diagnostic services.

A recent survey of American medical schools revealed that formal education in the principles of laboratory medicine is very limited or even nonexistent in many schools.[17]

The laboratory staff should also participate actively in the hospital's point-of-care testing program, including program design, testing and quality assurance.

THE FUTURE OF CLINICAL LABORATORIES

With the mandate to control health care costs and the movement to managed health care systems for the majority of the population, the market for clinical laboratory testing is becoming increasingly competitive.

Sharply reduced funding will be available for laboratory testing for patients enrolled in health maintenance organizations.

Consolidations and mergers among hospitals to increase their competitive advantage will affect the operation and organization of their respective clinical laboratories.

Under one scenario, acute-care hospitals in the future would maintain only "stat" laboratory capability, and would perform in-house only those tests whose results are required immediately for patient management. All other tests would be performed by a few very large centralized laboratories. In order to survive, smaller clinical laboratories must be able to meet the challenge by providing better laboratory service at a lower cost.

VI. REFERENCES

1. Letsch SW, Lazenby HC, Levit KR, Cowan CA. National health expenditures, 1991. *Health Care Financing Review* 14(2):1-30, 1992.
2. Helbing C. Hospital insurance short-stay hospital benefits. *Health Care Financing Review,* 1992 Annual Supplement:55-96, 1993.
3. Shimauchi A. Japanese Healthcare System: *In vitro* diagnostic tests and reimbursements. *Clin Chem* 40:1663-1667, 1994.
4. Muller B. Laboratory Diagnostics in light of massive changes in official health policies. *Clin Chem* 40:1658-1662, 1994.
5. Altman SH, Blendon RJ (eds.) *Medical Technology: The Culprit Behind Health Care Costs?* Washington, DC: Government Printing Office, DHEW Publication No. PHS 79-3216, 1979.
6. Galen MF, Gambino SR. Beyond Normality: *The Predictive Value and Efficiency of Medical Diagnoses.* New York, NY: John Wiley & Sons Inc., 1975.
7. National Committee for Clinical Laboratory Standards. *Cost Accounting in the Clinical Laboratory; Proposed Guideline.* NCCLS Publication GP11-T. Villanova, PA: NCCLS, 1993.
8. Saxena S, Wong ET. Does the emergency department need a dedicated stat laboratory? Continuous quality improvement as a management tool for the Clinical Laboratory. *Am J Clin Pathol* 100:606-610, 1993.
9. United States Department of Health and Human Services. Medicare, Medicaid and CLIA Programs: Regulations implementing the Clinical Laboratory Improvement Amendments of 1988 (CLIA '88). *Federal Register* 55:10896-10959, May 21, 1990.
10. Neumann BR, Suver JD, Zelman WN. The Capital Investment Decision. *Financial Management: Concepts and Applications for Health Care Providers.* National Health Publishing, Rynd Communications, 1984, pp. 351-369.

11. Arkin CF. Questions and Answers, *CAP Today* 6(9):70-1, 1992.
12. Palereti G, Maccaferri M, Manotti C, Tripodi A, Chantarangkul V, Rodeghiero F, Ruggeri M, Mannucci PM, on behalf of C.I.S.M.E.L. Fibrinogen Assays: A collaborative study of six different methods. *Clin Chem* 37:714-719, 1991.
13. Daley GQ, Van Etten RA, Baltimore D. Induction of chronic myelogenous leukemia in mice by the P210 bcr/abl gene of the Philadelphia chromosome. *Science* 247:824-830, 1990.
14. Ayscue LH, Ross DH, Ozer H, Rao K, Gulley ML, Dent GA. Bcr/abl recombinant DNA analysis versus karyotype in the diagnosis and therapeutic monitoring of chronic myeloid leukemia. *Am J Clin Pathol* 94:404-409, 1990.
15. National Committee for Clinical Laboratory Standards. Selecting and Evaluating a Referral Laboratory. NCCLS Publication GP9-T. Villanova, PA: NCCLS, 1991.
16. Emmanuel, L. Tracking your lab finances. *CAP Today* 8(5): 1-36, 1994.
17. Gottfried EL, Kamoun M, Burke D. Laboratory medicine education in United States medical schools. *Am J Clin Pathol* 100:594-598, 1993.

CHAPTER III
COST CONTAINMENT IN THE CLINICAL MICROBIOLOGY LABORATORY

James I. Mangels, M.S.

Stanford University Hospital Laboraotries
Stanford, CA

I. INTRODUCTION

For many years the issue of rapidly increasing medical costs was ignored by the health care industry. Most medical providers worked and trained in an environment of cost-based reimbursement–a system that offered few incentives for cost-effectiveness. Now, however, as the medical care system moves to prospective reimbursement, medical providers must address the issue of the correct use of resources and its relationship to the quality of laboratory results and patient care.

Laboratory and X-ray charges currently account for 25-30% of all hospital and ambulatory costs, and are increasing at an annual rate of 15%, making these charges the most rapidly increasing component of the nation's health bill.[1] Clinical laboratory services, therefore, are considered to be an important contributor to the general inflation in medical and health care costs. The magnitude of these cost increases has caused Federal agencies considerable concern, prompting them to develop policies to regulate clinical laboratories and control unnecessary costs associated with laboratory testing. The impact of Congress "CLIA 88" legislation, which restructures reimbursement practices, has forced health service organizations to consider more cost-effective management techniques for the clinical laboratory.

Two factors in particular contribute to the rising cost of laboratory services: increased test cost and increased utilization. It has been determined that one half of the laboratory cost increase is the result of increased costs to perform the test, and the remainder is due to increased utilization and new services, not due to inflation.[24,26] Therefore, changing utilization and ordering practices without compromising quality are an important aspect of cost-effectiveness in clinical laboratories. Authorities believe that 20-60% of laboratory tests may be unnecessary, and do not contribute to improved patient care.[26] Many erroneously assume that high cost and quality are synonymous—this attitude needs to be changed.

Health care in the 1990's is truly different from what we have known before. Some hospitals have managed care (medical insurance plans which control cost), some hospitals need discounts to compete, and some hospitals have high patient census, while others do not. The change to managed care and HMO's is coming about rapidly, and many laboratory personnel feel that their control of the laboratory is eroding. Currently, laboratory test pricing can be set on per diem rates, discounts, special packages, and/or capitated contracts. Today, the laboratory is no longer considered a source of revenue for the hospital. In addition, there has been a shift over the last few years from inpatient hospitalization activities to the outpatient setting, resulting in decreasing inpatient and increasing outpatient laboratory revenue. Although the test volume may be high, the revenue may be lower because of the need to have outpatient pricing competitive. At Stanford University Hospital, for example, there has been minimal growth among the inpatient population over the last five years, but an outpatient growth of 10-12% per year (Stanford data). These changes in patient population in hospitals will change our need for personnel, space, and capital.

Conflicts will inevitably develop in our changing environment as to what comprises quality health care, as opposed to cost-efficient health care. Regardless of how this conflict is resolved, laboratory testing that is clinically relevant will be essential and mandatory. Exhaustive microbiology (workup beyond that which may be needed for patient care), for example, is not clinically relevant or cost-effective, and practices such as this will no longer be permissible in the changing hospital economy. Not only must health care professionals develop the skills to manage hospital laboratories in the ways viewed to be most beneficial, efficient, and cost effective by the administration and the insurance companies, but they will need to provide laboratory testing that is clinically relevant.

II. ROLE OF THE CLINICAL MICROBIOLOGY LABORATORY

The role of the clinical microbiology laboratory is to 1) provide a diagnostic laboratory service, 2) help prevent the spread of infection, 3) provide data on specimen collection and transport, and 4) act as a resource to the medical staff by providing and interpreting data including antibiotic appropriateness. It is essential that laboratories provide accurate, high quality, and rapid testing so that the results are clinically useful. Responsibilities of the clinical microbiology laboratory also include ensuring the correct ordering, collection, and transmission of clinical specimens to optimize clinical relevancy, in addition to the conventionally accepted tasks of processing, reporting, and interpreting clinical specimens and results. Clinically relevant microbiology requires good communication between physicians and microbiologists to determine what the term relevant means in their institution.

The role of the clinical microbiology laboratory within a medical setting has changed. Whereas its previous role was as a generator of revenue paid on a test basis, it is now just another consumer of resources, and consequently will be one of the major targets for reduction of operation costs. Exhaustively good bacteriology produces irrelevant information, which misleads physicians into erroneous diagnosis and inappropriate therapy, and is much too slow for clinical application.[3,4,6]

The overall strategy for clinical microbiologists must be to play a leading role in the hospital's attempt to control costs without sacrificing quality. There is a need to deliver microbiology service of the highest possible quality, in addition to understanding cost and payment share.

Otherwise, arbitrary decisions may be imposed by administrators or insurance companies on the laboratory's budget. Reducing the cost of clinical microbiology services, without compromising the quality, requires consideration of the clinical relevance of all microbiological services. The microbiologist must be proactive rather than wait for decisions to be imposed on the laboratory. It is essential that administrators and laboratory personnel look at the impact of laboratory costs in relation to the "big picture", which is whether performing or not performing a laboratory test can shorten patient stay, and thus reduce overall hospitalization charges.

III. WHERE COST SAVING CAN BE EFFECTED

Cost control for the clinical microbiology laboratory can be achieved in three major areas. These are described briefly below and then discussed further in this chapter in detail.

A. Factors related to patient testing include what test or tests should be requested and how often. Are there tests not requested by the clinician that actually should be ordered? The issue of utilization control is crucial to prevent over utilization as well as under-utilization of clinical microbiology tests. In addition, there are other factors related to patient testing that must be evaluated for cost, such as, how should the results be used, and how should the results be communicated to the physician for prompt use?

B. Factors related to laboratory technical operations should be examined for cost savings. What processing, collection, transportation, ordering, and reporting methods are being used? How extensively should the laboratory work up a specimen. How rapidly and by what methods? In addition, the quality control procedures and their frequency should be examined, as well as what susceptibility testing is deemed necessary. The laboratory's role in controlling antibiotic use by the physicians through reporting practices, data collection, and antibiotic review committees should be considered.

C. Lastly, factors related to laboratory management operations need to be examined for cost improvement. Are laboratory resources (personnel and equipment) being used properly? Is the skill mix adequate for the laboratory? Does the laboratory know and monitor its current labor and supply costs? Does the laboratory use automation, and is it being utilized properly and effectively? Is laboratory productivity being monitored and compared, and is it being compared to a reasonable standard, or against other similar medical institutions? Are computers used appropriately in the laboratory? Should some laboratory work be referred and sent out to an outside laboratory? Should some of the tests that are currently being sent out be brought back into the laboratory?

FACTORS RELATED TO PATIENT TESTING

Since about half of the overall increase in health care spending is the direct result of increased utilization of medical services, a change in utilization procedures will greatly affect cost effectiveness. The greatest cost-efficiency will be gained from changing utilization practices.[4,24,31,36]

Changing ordering practices and utilization are some of the most controversial and clearly some of the most difficult areas for the laboratory to alter since many aspects of such a program cannot be controlled by the clinical laboratory. It has been suggested that the place to start is with the physicians in training, rather than with currently practicing physicians.[10,13,19,22] It may be easier to change the behavior of interns, residents, and fellows by performing audits and in-service training, when they are employed by the medical center rather than when they have their own practice. Most practicing physicians acquire knowledge on a random and irregular basis, from continuing medical education courses and consulting sub-specialists' recommendations. If an algorithm to change utilization of the laboratory is to succeed, the impetus for this development must be envisioned by all participants as a cooperative educational venture. The educational design should be informational, not punitive. It is imperative to get physicians' participation and involvement in the development of laboratory testing algorithms and not to allow these to be totally dictated by the laboratory. Table 1 provides some cost-effective tips for the clinical microbiology laboratory.

Table 1
Cost-effective Tips for the Clinical Microbiology Laboratory

- Elimination of check-off boxes from laboratory requisition form
- Screening to check specimen quality
- Rejection of inadequate and replicate specimens
- Elimination of terminal subculture on radiometrically-negative blood cultures
- Identification of common pathogens by simple, rapid methods, including colonial and/or microscopic morphology or direct fluorescent antibody staining.
- Limiting identification and susceptibility testing to two organisms per specimen
- Reduction in the over identification of certain isolates, e.g., speciating coagulase-negative staphylococci or anaerobic cocci from all specimens
- Elimination of telephone reports before 10 AM
- Elimination of many interim reports.

The success of the program should be judged by its ability to improve test ordering habits.[31] The primary objective should be to arrive at a diagnosis as soon as is reasonable, using the most specific, efficient, and cost-effective diagnostic methods. The utilization program developed should have educational value, and be flexible enough so as not to restrict the clinician. Some laboratory diagnostic information may affect decisions on whether to obtain additional confirmation from the microbiology laboratory. Many hospital teams (physicians, nurses, ancillary personnel) are currently developing clinical pathways that suggest policy for specific laboratory testing at certain stages depending on diagnosis and/or clinical indications.[36] These clinical pathways for each diagnosis will become the standard of care in the future.

What is needed is a broad view of laboratory costs as they relate to the whole hospital. This view should incorporate issues of patient management, utilization, and clinical relevancy in order to determine overall cost. This is an essential link to make, but often it is a difficult one for laboratories to make by themselves. What is often needed is assistance from administrators with a clear understanding of hospital costs, and a vision of the role of the laboratory as a vehicle to shorten patient stay and provide improvements in patient care, rather than the administrator's view of the laboratory as an adversary, and an expense to patient management. In a managed care environment, the laboratory is no longer considered a revenue center; it is a cost center. Whether the laboratory can generate revenue is no longer important. To survive and thrive in such an environment will require microbiologists and administrators who think not just in terms of the laboratory, but of the whole care system, "the big picture", so that they can determine whether a test, or procedure is cost-effective.

In the past, clinical laboratory accounting practices have had a profound effect on the development and management of clinical laboratories because revenues were highly leveraged relative to true laboratory costs, and were essential to the fiscal health of a hospital. However, such practices lead to dis-incentives for the implementation of more economical use of laboratory services. There is a price/cost disparity, which can be traced directly to hospital budgeting methods, that uses the clinical laboratory as a profit center to support unrelated deficit-producing hospital operations.

To achieve meaningful cost-containment, the microbiology laboratory will be required to adopt test utilization patterns, or to act with clinical services, to restructure therapeutic strategies based on laboratory test results. Success in this new out-of -laboratory arena will be difficult to achieve. Protocols will be difficult to structure because of the habitual test use patterns of most providers of primary care.[35] Unnecessary tests are routinely performed that provide no benefits to the patient or the treating physician. These tests may be ordered for a variety of reasons. Often the main concern is to prevent litigation. Protocols once instituted are often hard to control, and periodic education of staff members may be necessary. Good laboratory orientation programs are a must for competent, cost effective clinical microbiology.

Often a question arises as to how test ordering can be reduced in the face of defensive medicine. The physician is pressured to order every possible test to protect against a lawsuit in case of an adverse outcome. The answer may come from the development of clinical pathways that guide testing practices and therapy. Successful efforts to reduce the misuse of laboratory resources include long-term commitment to cost containment, and a combination of educational programs with active intervention to reinforce these educational efforts. The concept of cost containment alone cannot be used as the sole reason for a cost containment program. Instead, it is important to emphasize the improvement in the quality of care that will occur as a result of reducing over-utilization, under-utilization, and mis-utilization of laboratory tests.

General Considerations

A number of factors must be considered to determine which patient testing services the laboratory should provide. Factors to consider include the following: the patient population that the laboratory is serving (outpatient vs inpatient), the culture mix, the specific type of patient mix (illness), volume of tests performed, available instrumentation, available computer support, labor and supply costs, orientation to rapid results, accuracy, hours of operation, and Federal and State Regulations.[24,30] Some of these factors will be discussed below.

Strategies for Cost Effective Microbiology by Altering Patient Testing Services

- Eliminate misuse and overuse of laboratory tests by setting a policy stating which tests should be requested and how often.
- Eliminate unnecessary tests.
- Determine whether there are tests not requested by the clinician that should be ordered.
- Use cost saving technology (automate when appropriate).
- Increase physician awareness of the value of appropriate diagnostic testing.
- Use computers for prompt and accurate laboratory reporting, and to provide information on specimen collection and frequency.

Utilization of Laboratory Testing

The practice of standing orders for routine laboratory tests has been shown to greatly increase laboratory utilization without considerable improvement in patient care.[5,9,31] Hospital wide teams that examine standing orders can provide realistic expectations of laboratory services and reduce frequency of laboratory testing. A hospital-wide or laboratory computer is essential to properly evaluate a reduction in volume of standing orders. A good computerized information system can help relieve concerns about compromising the quality of patient care. For instance, physicians may be willing to abandon the practice of daily testing in the absence of changes in test values if the physicians are convinced that daily updates will be provided, and that all clinically significant changes will be brought to their attention.[30,31,36] Table 2 provides examples on suggested limits on specimen collection.

Table 2
Suggested Limits on Microbiology Specimens

Specimens	Limits
Urine, stool, expectorate sputum, wounds	One per day, not to exceed three per week
Upper respiratory	One per day, not to exceed three per week
Blood	Three per day, not to exceed six per 48 hrs.
Stool for ova and parasites	Three per patient admission or illness
Specimens for viral isolation	One of each type per patient admission
Specimens for AFB	Three of each type per patient admission
Urine or CSF for acid-fast bacilli	Done following infectious disease consult
Serum bactericidal test	Done only following ID consult
Specimens obtained by invasive procedures	No limit

Microbiologists and clinicians may have different perceptions of the sort of laboratory testing that is rational. In the clinician's view, a good microbiology test is one which provides useful clinical information quickly. It is often irrelevant to the clinician that a test is labor-intensive, or requires huge outlays of equipment or supplies. If the clinician believes this is the only way to make a diagnosis, then to the clinician the test is justified. The marked difference in clinician's and microbiologist's perceptions and attitudes is an important factor when changes are made in microbiology services. The focus must remain on performing tests as accurate as possible, and at the least cost, rather than on cutting costs regardless of quality and type of laboratory services needed for patient care.

Physicians may request unnecessary tests because of their insecurity in establishing a diagnosis, or for academic curiosity, or because of poor turn-around time of the laboratory.[5,26,31] Refusing improper requests and rejecting poor quality specimens should now be thought of as providing a needed service to the physician and to the patient. Excessive numbers of blood cultures, urine cultures, daily sputum cultures, daily cultures of various drainages, or multiple wound, or superficial cultures are all examples of overutilization of the microbiology laboratory.[4,15,26]

The microbiologist needs to discuss proposed changes and to implement these changes with the involvement of the clinical staff. The best approach is to show how much better patient care can be given with the proposed changes. Such changes may be to modify tests that are already being performed, or to discontinue some low-volume labor-intensive tests. In addition, one might examine whether the microbiological work-up can be limited to something short of full microbiological evaluation. When new tests are considered for the microbiology laboratories, they should be examined for benefit and cost.

Continuous effort by the clinical laboratory is necessary to control incoming work load and to control patient testing factors. Many laboratories do not have the personnel resources needed to implement and maintain a work load control mechanism. On the other hand, many laboratories do not want to encounter the negative feedback that a critical approach to altering test requests can generate. Some physicians assume that all test requests are clinically necessary even though existing data do not support this premise, and other physicians believe that whatever test they request should be performed without question.[12,15.19,31]

The concept, that clinicians submit specimens and laboratories run the requested tests without question, no longer holds true in this era of "cost accountable-patient outcome clinical laboratory testing". Specimen quality must be evaluated prior to plating and subsequent workup. In addition, patient parameters and other test results may indicate that certain test requests are inappropriate, e.g., latex antigen tests on CSF when the specimen contains normal levels of protein, glucose, and no cells. The clinical laboratory must now evaluate both the patient and the specimen to determine the most appropriate specimen procedures, and to determine which specimens and tests will be performed and which ones will be rejected.

Examples of How to Control Patient Testing Factors

1. Modify tests already being performed. There might indeed be some low-volume labor-intensive test than may be targeted for discontinuance. In addition, one might examine whether the microbiology work-up can be limited to something short of full a microbiology evaluation.
2. Assess whether or not a particular test has sufficient diagnostic value for a given diagnosis, and establish criteria to limit specimen processing to tests likely to be clinically relevant.
3. Eliminate unnecessary, inappropriate tests. This will also reduce the production of clinically misleading information and information overload, improve the quality of care, and reduce costs.
4. Examine new tests carefully before including them in the clinical microbiology laboratory repertoire. Consider three specifics: benefit, clinical relevancy, and cost.
5. Monitor specimen quality to help control patient testing factors and improve quality. Specimens that are not collected or transported properly, even when handled optimally within the laboratory, are likely to provide misleading results. Protocols to assess specimen quality will be welcomed by the clinician.
6. Consider eliminating tests that have a low clinical significance, such as a request for Mycobacterial cultures from spinal fluid. Similarly, consider suggesting tests that are not currently being requested, such as anaerobic cultures on abscesses obtained from surgery. Table 1 may also provide some useful tips.

Education to Change Laboratory Requesting Practices

Changing laboratory practices, to either add a new test or delete current tests, requires an understanding of physician and patient care needs. Changing testing practices without educating the medical staff is subject to failure. Physicians who do not understand the rationale and procedures for a changed test, may cost your laboratory staff time and money in delayed explanations, repeated tests, and stressful interactions. It is best to educate the entire staff- physicians, nurses, and laboratory personnel- before initiating change.

In-service presentations to small groups of physicians at lunch are particularly effective.[19] Announce these sessions well in advance to as many physicians as possible by posting flyers, sending electronic or voice mail messages, and adding the announcement to physician newsletters, or in-house newspapers. The date, time, and location should also be announced at physician staff meetings. Schedule at least two identical presentations, preferably on different days of the week, and at different times of the day, to accommodate physicians' schedules. Teaching tools such as slides and handouts work well; physicians are accustomed to this format. Pass around and demonstrate any specimen containers, or other new products. Above all, provide the rationale for the changes you propose to make. Emphasize the cost-efficiency, ease of physician use, and benefits to patient care of the new or revised procedure.

Similarly, in-service training for nursing personnel and for the laboratory staff is crucial for the success of bringing about a positive change. Many of the procedures described for physician in-service can be used for the nursing staff and for laboratory personnel. It is important to accommodate their schedules; go directly to the unit for nursing personnel, and plan programs at different times to get cooperation. Obtaining the buy-in and cooperation of the nurses is vital. It is important that they feel part of the solution rather than being dictated to by the laboratory.

Finally, the buy-in of your own laboratory staff is vital. If they do not understand the significance of the proposed changes, they may undermine the process of change. In-service training in small groups over lunch or breaks works well, particularly if you provide some treats. You will need to provide the training at many different times for the program to be successful. Most of all, the staff needs to feel that they are listened to, and are part of the solution to improve patient care.

LABORATORY TECHNICAL OPERATIONS

The most conventional approach to cost reduction is simple straight cost cutting of laboratory technical operations and of the operational budget.[3,9,12,15,30] However, these approaches may affect quality and service if you let them. Examples of altering technical operations would include such things as reducing supply costs, limiting the extent of workup and identification by the clinical microbiology laboratory, and redistributing and combining the work into fewer workstations.

Straight cost cutting within the laboratory does not require alterations or changes in current test ordering behavior. It does require, however, a close scrutiny and careful cost accounting of laboratory expenses.[20,23,30] These may include changes in supply vendors, service contracts, better use of reference laboratories, elimination of redundant proficiency testing, extensive quality control and quality assurance practices, and freeing technologists from tasks that can be performed by technicians, clerks, or laboratory assistants.

Supply cost represents approximately 9 to 20% of all direct costs of the laboratory.[3,12,24,26,30] This expense can be improved by group purchasing contracts, improved inventory control , persuasive negotiation with vendors, and volume discounts. Buying economical reagents and preventing waste can help, but the greatest cost savings in supplies comes about by group discounts--forming consortia with other laboratories.[30,35]

The greatest expense of the clinical microbiology laboratory, however, is in labor.[24,26,30] Microbiology is an expensive laboratory service because it is labor-intensive, and technologists' salaries generally account for 50 to 70% of a laboratory's operating budget.[26] Even in this day of automation, most microbiology procedures and methods are performed manually. Microbiology has not been as automated as chemistry or hematology, so it is still labor-intensive. Therefore, methods of reducing the labor cost of the clinical microbiology laboratory are essential to reduce overall cost. For example, a medium-sized microbiology laboratory may produce about one-quarter of the number of billable tests per full-time equivalent as compared to a hematology or chemistry laboratory.[24,26] With more automation, and with new molecular biology methods, microbiology will begin to resemble the chemistry lab, and in time, the same level of productivity from microbiology will be expected as from the hematology or chemistry labs.

A laboratory can reduce overall personnel expenses by hiring lower-qualified personnel, and thus change the skill mix. These measures can reduce direct expenses by a minimum of 20%.[24] Other factors that reduce labor costs are eliminating excess quality control, reducing hours of operation, evening and night shift services, reducing stat tests, and cutting down on the overall services available. The impact of these service reductions on patient length of stay needs to be determined prior to initiation.

Other practices such as reducing the frequency of testing and increasing the batch size, will reduce labor and materials costs as well as certain overhead items. However, some microbiological tests cannot be delayed for technical reasons. Limiting stats to certain procedures, and allowing only emergency or acute care units to order such tests without specific approval can also be effective.

The following methods to reduce costs by changing internal operations will affect total labor cost, and should be explored by the clinical microbiology laboratory. Some of these changes may be invisible or transparent to the physician user, others may not.

Methods of Cost-Reduction of Laboratory Technical Operations

1) The requesting process
2) The quality of the clinical specimen
3) Collection, storage, and transportation of clinical specimens
4) Limitations on testing, specimen frequency, and rejection criteria
5) Extent of primary set up
    a) screening procedures
6) Extent of workup and identification
    a) Examples of cost-efficient workup and identification
7) Extent of susceptibility testing
    a) Role of the laboratory
    b) Controversies with susceptibility testing
    c) The use of automation for susceptibility testing
8) Automation and instrumentation
9) The use of laboratory information systems
10) Alternatives to conventional culture techniques
11) The clinical microbiology laboratory in the big picture
12) The use of reference laboratories

The Requesting Process

Often the initial requesting process can lead to inappropriate and costly clinical microbiology practices.[4,30] The initial requesting process is the ideal place to let the physician and other medical staff know laboratory policies and guidelines for specimen submission, frequency, and transportation. There is a need to have a requesting system that provides clear definitions of what information is needed from the requestor and exactly what specimen is required by the laboratory. For example, the request for a generic "wound culture" is not satisfactory. Not only does it fail to provide the microbiologist with adequate culture information, but also it does not allow the physician to appropriately evaluate the culture results. What is needed is a notation of the exact specimen source and location so that the specimen may be cultured appropriately. The laboratory needs to immediately follow-through if the information provided on the requesting form is not complete. The ideal situation is to have a remote requesting computer terminal that prohibits the completion of the form until all information is provided.

The following are examples of useful information to be given to the physician or nurse before they collect and request specimens for cultures:

1. Specimen collection guidelines, containers to use, patient instructions where appropriate, and transport mechanisms.
2. Rejection criteria, how the physician will be notified if the specimen is not satisfactory.
3. Reporting guidelines, interim status, and final reporting practices.
4. Expected turn-around-times, hours and times of testing.
5. Cost

The Quality of the Clinical Specimen

Physicians often require guidance on the most appropriate specimens: how to collect them, the frequency of their submission, and methods of ensuring specimen quality.[3,4,5] Also, physicians are often unaware of the effect of contamination with indigenous microflora on specimens such as blood, urine, sputa, and wounds. Therefore, the microbiologist needs to provide information to the clinician about contamination. The microbiologist needs to institute procedures for the rapid delivery of specimens to the laboratory. Finally, the clinical microbiology laboratory needs a clear statement of policy, reinforced to all the medical staff, to prevent any misunderstanding on the transport and collection of specimens so as to prevent poor specimen quality.

Monitoring specimen quality can help reduce cost to the laboratory and to the medical institution. Specimens that are not collected or transported properly, even when handled optimally within the laboratory, are likely to provide misleading results. Improved specimen quality can help prevent the physician from acting on incorrect or irrelevant data. Physicians will welcome protocols to assess specimen quality if they understand the consequences of poor quality. The jargon that is often used in industry and among computer programmers is "garbage in--garbage out". The same idea holds true for clinical specimens as well. If the specimen initially is of poor quality, the culture results will also be of poor quality. Laboratory costs can most effectively be controlled by more appropriate test requests , higher specimen quality, and more timely reporting.[4,5,6,13,15] These steps will have probably have more impact than anything else in reducing costs in the clinical microbiology laboratory.

Collection, Storage and Transportation of Clinical Specimens

Proper collection is essential for the successful recovery of the causative microorganism(s) by the clinical microbiology laboratory.[7,9,12,30] It is clear, however, that the microbiologist must take an active part in this process to provide adequate supplies and proper instructions to ensure correctly collected samples. In-service programs to nurses and physicians are essential. Guidelines on proper techniques should be available to the medical staff to prevent contamination, and to avoid inclusion of normal microbial flora found throughout the body. Often the microbiology laboratory is asked to perform cultures on specimens that are contaminated, adding only to the over-all cost to the laboratory. Frequently contaminated specimens include: sputa, urine, superficial wounds, anaerobic cultures, and blood cultures.[7,16]

Improper storage of specimens can also result in unnecessary laboratory testing and certainly in erroneous reports. For example, a urine sample with just a few colony-forming units (CFUs) of bacteria left out at room temperature could easily yield a colony count that may be considered significant. This may necessitate full identification and susceptibility testing, and possibly lead the clinician to unnecessary and/or incorrect therapy. This adds not only to laboratory costs but to overall hospital cost.

The inappropriate storage of sputum specimens can result in the normal respiratory microbiota overgrowing potential pathogens and yielding misleading information. The sputa specimen may initially meet acceptable microscopic screening criteria, but potential pathogens may be masked by the overgrowth of commensal microorganisms.

Unfortunately, the transportation of specimens from nursing stations to the laboratory is often left to individuals who often are never clearly instructed in the significance of their responsibilities. In addition, unnecessary delays and improper storage lead not only to needless workup by laboratory personnel, but may delay appropriate therapy, prolong hospitalization, and therefore add expense to the institution.

Laboratory personnel must insist that the individuals obtaining specimens be thoroughly familiar with established collection and transportation procedures. They must be aware of the proper transport containers to be used for a particular specimen, the proper time for collection, and the proper method of storing specimens prior to and during transport.

Proper collection of specimens is also imperative, both from the cost cutting point of view and the patient's welfare. For example, anaerobes from contaminated areas, and Foley catheter tips submitted for the diagnosis of a urinary tract infection are inappropriate, and should not be accepted. The medical and nursing staff need to be instructed on the proper method of disinfection of skin areas. The isolation of any microorganisms from normally sterile body fluids requires complete identification, and when appropriate, antimicrobial susceptibility. The presence of skin contaminants will lead not only to wasteful efforts on the part of the laboratory, but may lead to unnecessary and/or inappropriate therapy.

It is imperative that the laboratory follow-through on collection , transportation, or testing frequency problems immediately, to provide corrective in-service training while the problem is still fresh. It is often best to work with the cooperation of a nursing in-service leader or nursing manager. If they understand what the patient consequences are, they can often act as a resource in behalf of the laboratory since they usually see the physician much more frequently than anyone from the laboratory.

Limitations on Testing, Collection Frequency, and Rejection Criteria

It is essential that the laboratory sets limits on testing and specimen collection frequency, and have clearly defined rejection criteria immediately available for the medical and nursing staff.[4,7,26] Often, laboratories receive an unwarranted number of repeat or duplicate specimens within the same day, or within a short period of time. Excessive number of blood cultures, daily sputum cultures from patients without signs of pneumonia, daily cultures of various drainages, or cultures of superficial wound material are some examples of over utilization of the microbiology laboratory. These specimens rarely contribute useful information for patient management. Although multiple factors contribute to excessive use of laboratory tests, physicians are the single most important factor in the expansion of laboratory testing. Again, in-service education of the medical staff may be one of the most important mechanisms in implementing effective change in laboratory testing practices.

Because of limited information on the status of the patient, it is sometimes difficult for the clinical microbiologist to set limits on the total number of specimens required for a diagnosis. On the other hand, the physician may not know that multiple samples have been submitted, and may be very receptive to assistance on the recommended number of specimens to be collected within a certain time period. The policy of limiting specimens should be undertaken only after careful study and discussions with clinicians to ensure that no negative effect on patient care will result.

Some general guidelines for the clinical microbiology laboratory are as follows: a single sputum, urine, stool, or throat specimen collected with a 24-hour period for microbiological analysis is appropriate. Daily samples without clinical change of the patient do not provide added clinical information to the physician, but clearly add to the overall cost. Three early morning sputum specimens collected on successive days are sufficient for the diagnosis of the majority of pulmonary mycobacterial infections, and once the laboratory diagnosis of a mycobacterial infection has been established, follow-up specimens should be limited to one specimen per week while the patient is undergoing antimicrobial therapy.

For blood cultures, three blood cultures collected within a 24-hour period have been shown to permit the diagnosis of almost all cases of septicemia.[7,14,16]

The data from the culture of intra-vascular catheter tips (Hickman lines) has been shown to be of little clinical significance. Many endotracheal culture requests are useless; such specimens should be first evaluated by Gram stain. Endotracheal specimens may be cultured only if microscopy shows organisms and fewer than 10 epithelial cells per low-power field.

For the diagnosis of an intestinal protozoan infection, three stool samples collected on alternate days is usually sufficient to establish a diagnosis. Studies have shown that examining stool samples for ova and parasites is of no value in patients who develop diarrhea after being in the hospital for more than three days.[11,16,21,25] In our institution, half of all requests for ova and parasite examination are for patients beyond this period. Eliminating these unnecessary specimens results in a considerable savings without a loss of diagnostic information. Also, the likelihood that such a patient may have *Shigella* or *Salmonella* is very low.[11] Of greater significance for these patients is testing for the toxin of *Clostridium difficile*.

The culture of Foley catheter tip for the diagnosis of a urinary tract infection is inappropriate and should not be accepted.[13,16,26] Surface cultures of decubitis, anaerobic cultures of vagina, cervix, or oral sources, and often periodontal material are inappropriate for culture, and may even yield misleading information.[7,16,32] Cultures from decubitus ulcers obtained with a swab are inappropriate and should never be processed. Since many anaerobic infections arise from areas that are contaminated by mucosal microbiota, sputa, bronchial washings, brushings, and vaginal secretions, they should not be accepted for the purpose of isolating anaerobic bacteria. Table 3. provides an example of tests that are appropriate and inappropriate for anaerobic culture. The presence of skin contaminants from superficial wounds will lead not only to wasteful efforts on the part of the laboratory, but the information reported may lead to unnecessary and/or inappropriate therapy.

Table 3.
Appropriate and Inappropriate Specimens for Anaerobic Culture

Appropriate	Inappropriate
Expectorated sputum specimens	Surgically obtained
Bronchial washings	Deep abscess
Small bowel contents	Aspirates
Voided or catheterized urine	Sterile body fluids
Vaginal or cervical swabs	Biopsy material
Female genital tract cultures, except by needle and syringe	Tissues
Surface swabs	

The clinical microbiology laboratory needs a clear statement of policy, communicated to the medical staff, to prevent any misunderstanding on the collection and acceptable number of specimens per day. Written specimen collection guidelines and clearly defined protocols on the rejection criteria are essential for cost-effective clinical microbiology. In addition, there is a need to have a practical mechanism for alerting physicians or nurses immediately when problems arise. Clearly written instructions on specimens collected by the out-patient department are also essential to reduce expense. This will eliminate confusion to the patient on specimen collection, and will prevent the need to submit another specimen.

Clearly, laboratories need to become proactive in curtailing useless tests, duplicate samples, samples that have been delayed in transient, and/or placed in inappropriate containers. The physicians must always be kept well informed of laboratory policy so that special needs will be accommodated, and patient care will not suffer when tests are limited to save money. In some circumstances, what appears to be a costly procedure, may, in fact, prove cost-effective in the overall achievement of better patient care. Some circumstances may dictate a compromise between what is requested and what the laboratory may be able to perform.

In the past, clinicians often have grown accustomed to the idea that the laboratory will perform, without question, all the tests they request. Efforts, therefore, to set limits on testing practices and to discuss possible changes often come as a surprise, and sometimes with disagreement. These may be interpreted by the physicians as a challenge to their responsibility for patient care. These discussions are not intended to result in a conflict, however, the idea of cost and appropriate laboratory testing has not been part of the dialogue in the past, and may be hard for the physician to accept. In the future, it is likely that the clinician will find further restraints imposed on laboratory testing and other elements of patient care by hospital administrators and third-party payers. It must be emphasized that guidelines on the frequency of specimen submissions are subject to modification on a patient-by-patient basis, and that this requires a continuing dialogue between the clinician and the laboratory.

Cost-effective microbiology also requires the elimination of tests that are not helpful to the patient. To effect all these policies and changes, it is essential that good communication exits between the laboratory and clinician. Therefore, providing education on specimen quality and testing frequency is one of the most important functions that laboratory personnel can and must provide.

Ultimately, when unnecessary tests are eliminated, cost reduction results. Whether a particular test is necessary is a clinical judgment, but there are many existing abuses. The problem is in convincing physicians to stop requesting the unnecessary tests. Changing physician concepts of how to manage infections can produce a decrease in laboratory testing. For example, the development of clinical pathways, which reduce utilization, limit specimen submission, and control the extent of processing and identification, and provide useful information, limit work on sub-optimal specimens by encouraging the medical cost/effectiveness.

Extent of Primary Setup

A significant amount of the laboratory's resources are directed toward the primary setup and workup of cultures, namely the isolation, identification, and susceptibility testing of clinically significant microorganisms.[5]

The initial microscopic analysis of clinical specimens is a cost effective procedure that rapidly provides important information. If used appropriately, the Gram stain can serve as an initial guide for therapeutic intervention, and can provide information for the selection of appropriate media for the isolation of potential pathogens.

Our laboratory uses a Cytospin which has greatly increased the sensitivity for detecting low numbers of microorganisms from body fluids. The implementation of automatic protocols by some laboratories in which a number of laboratory procedures are performed automatically on a specimen, usually dependent upon the source, may allow the detection of unsuspected potential pathogens. However, these protocols must be analyzed against clinical need, clinical relevancy, and over-all cost benefit. A simple decision is not always possible.

Information regarding the type and source of a specimen and the patient's history are useful in determining the type of media to be employed. For example, screening of stool specimens for *Vibrio* or *Yersinia* when the history of the patient does not suggest exposure (seafood or travel) may not be cost-effective.[11,17,21,29] The overall frequency of recovery of *Yersinia* is so low, that routine usage of special media just for this organism may not be necessary. Furthermore, in many instances, this organism may often be recovered from conventionally used media (MacConkey) so that special media or techniques are not warranted.[17]

Screening Procedures

Microbiology personnel's time may be best used studying the colonial morphology of microorganisms on routine media.[2,4] In combination with microscopic morphology and origin of the specimen, this information allows laboratory personnel to separate the normal flora from potential etiologic agents of an infectious disease. The financial consequences of this knowledge can be very significant since it will eliminate unnecessary microbial identification.

Tremendous cost savings can be achieved if specimens unlikely to yield diagnostic information are not processed beyond initial evaluation. For example, the screening of expectorated sputum for the presence of squamous epithelial cells, with or without polymorphonuclear neutrophils, is an effective way to improve the quality of these specimens and to prevent unnecessary work.[5,7,16,25,35] Another idea is to screen material from wound cultures for the presence of polymorphonuclear neutrophils which indicates an infectious process; the presence of squamous epithelial cells indicates surface skin contamination.[5,7] There are obvious limitations, but in the majority of cases this can provide guidance.

Although there is a lot of controversy regarding the screening of urine samples prior to culture, some laboratories have shown cost savings as well as improvements in patient care by initial screening of urine samples.[7,8,35] Physicians will benefit by receiving negative results at least 1 day sooner than by conventional culture, which aids in preventing antibiotic overuse, and allows earlier evaluation for additional causes of urinary tract related symptoms. New screening procedures are being developed for many other specimens from infection sites. *C. difficile* screening for presence of *Salmonella* or *Shigella*, *Legionella* screening from urine by latex agglutination, and *Giardia* EIA are just a few examples.

The use of screening procedures will ensure the workup and reporting of clinically useful information by eliminating the processing of poor quality specimens. Screening procedures will also reduce the cost associated with needless identification and antimicrobial susceptibility testing procedures. Bartlett[5] showed a cost reduction of 19% in his laboratory when screening procedures were instituted and followed, compared to processing all clinical specimens regardless of quality and without any processing control policies.

Extent of Work Up and Identification

The extent of workup and identification of microorganisms by the clinical microbiology laboratory must be appropriate to the specimen type, the resources of the laboratory, and to the needs of the physician who ultimately is responsible for the patient.[3,5,15,35] However, the complete definitive identification of all organisms from all clinical specimens is probably not necessary from a clinical perspective, and wastes valuable laboratory resources. Some organisms recovered from certain specimens need only simple biochemical tests for identification so as to be clinically useful.

The initial question for each laboratory to determine and answer is how rapidly and how completely they need to identify bacterial isolates. Identification, of course, is warranted for isolates from sterile body fluids such as CSF and blood as well as for wound specimens collected by deep aspiration and/or submitted from surgery. It is probably an overkill, on the other hand, to completely identify all organisms recovered from urine, superficial wounds, and from expectorated sputum specimens. In many cases, isolates can be identified with an acceptable level of accuracy, by using the skilled technologist's observations of the appearance of colonies on certain media, and a few spot tests. Many clinical situations exist in which rapid, presumptive information is more important than delayed, definitive information.

One of the most cost-effective measures available to the microbiologist is to have available brief schemes to presumptively identify bacteria based on colony morphology and simple biochemical tests. For example, a *E. coli* can be identified with a specificity of >99.9%, based on growth and lactose fermentation on MacConkey agar, indole positivity, oxidase negativity, and Beta-glucuronidase positivity within one hour, at an average cost of twenty cents per isolate.[2,8,12,15,18,33,35] Likewise, a swarming Gram-negative bacteria recovered from a urine sample that gives a negative indole reaction can be identified as *Proteus mirabilis,* while a swarming Gram-negative rod that is indole positive may be identified as *Proteus vulgaris.*

For other members of the *Enterobacteriaceae,* biochemical identification is probably necessary unless one chooses to use some generic term, such as "enteric gram-negative rod," on organisms recovered from superficial sites. The greatest cost savings may be achieved with the presumptive identification of the *Enterobacteriaceae*, due to the frequency of their isolation and the wide variety of non-kit and non-instrument rapid commercial test products available. However if the laboratory feels that is imperative to biochemically identify each and every isolate completely from all specimens, the total cost for identification can get out of control. Bale and others[2] showed a savings of $200 per 100 *Enterobacteriaceae* identifications by using a limited identification scheme, such as growth on MacConkey, spot indole, and the use of additional biochemicals. Using this scheme, Bale showed that they were able to accurately identify 98% of the isolates at the genus level. The use of spot tests such as MUG or MUCAP, which look for fluorescent 4-methylumbelleferone esterase, can be extremely useful for the rapid and cost-effective identification of some species of *Enterobacteriaceae*.[33]

Other brief schemes may be used for a variety of other commonly recovered organisms from clinical specimens. A gram-negative, oxidase-positive rod which produces a green pigment (pyocyanin), has a characteristic grape-like smell, and grows at 42°C can be reported as *Pseudomonas aeruginosa* .[7,18,35] In most instances, the identification of staphylococci does not require more than a simple coagulase test which can accurately identify the most clinically significant member of this genus.

Although the presence of coagulase-negative *Staphylococci* has clearly increased in blood cultures, this may present an exception to this scheme, whereas in other sites, the use of only the coagulase test may be appropriate.[18,34] Other methods can be used to decrease the cost and time required to obtain identification. The PYR test, a simple biochemical screen, can be used for rapid identification of *S. pyogenes* and *Enterococcus* species. The latex or coagglutination reagents can be used for rapid serologic grouping for other beta-hemolytic streptococci, and the identification of *C. albicans* may be made if the yeast forms germ tubes within 3 hours at 35°C.[7,16]

Identification of organisms from superficial wounds may not need to be identified to its "mothers maiden name"; however, blood cultures or organisms recovered from spinal fluid specimens, or from normally sterile sites may require full identification and susceptibility testing. Another potential cost savings has been the increased attention for the identification of species of coagulase-negative staphylococci to a species level from urine or mixed culture wound specimens. For example, a number of studies have shown *S. saprophyticus* to be a significant cause of urinary tract infection, especially in young women. This may be important for studies of pathogenesis and epidemiology but does not necessarily mean this species must be separated from the other coagulase-negative staphylococci.[7,34,35]

In one study by Williams et al.,[34] *S. saprophyticus* was the most common staphylococcus causing symptomatic urinary tract infection. However, an equal number of cases was associated with other coagulase-negative species, and 40% of the *S. sapropyhyticus* isolates were not associated with disease. Therefore, a report of *S. saprophyticus* from urine may not be of any greater value to the clinician than one of coagulase-negative staphyloccocci. However, in spite of these arguments, it is often perceived by many that simple speciation is not enough, and that the reputation of a laboratory is directly correlated with the sophistication and the complexity of one's identification schemes.

Microbial identification beyond that required for diagnosis or therapy may gratify the ego of some laboratory personnel, but can involve additional cost and a delayed report.[2,5,26]. Although the laboratory may take pride in its ability to identify to species and subspecies, and to stereotype various organisms, there are times when such procedures become academic exercises rather than clinical tools. In many instances microbiologists have been just as guilty as clinicians of excessive zeal in pursuit of complete identification. Clearly, self-laboratory generated work, beyond that which is required and needed, can add to the over-all cost. The extent to which microorganisms are identified to the genus, species, subspecies, and type level must be related to the clinical relevance of the information. Good judgement is necessary from the laboratory staff to know when to apply the techniques to obtain detailed taxonomic information.

The availability of commercial kits and automated procedures for detailed identification have made it even more tempting to identify to species every organism isolated. If a clinical reason cannot be advanced for this practice, it is often said to have epidemiological value. New techniques such as DNA probes can significantly shorten turnaround times on certain specimen types, and provide detailed identification. These newer methods are often more expensive and labor-intensive than conventional methods, and the workload can significantly increase while specimen counts remain constant; this leads to increased operating costs and decreased productivity. The strategies that address cost containment should be evaluated for their utility in each laboratory.

The quantitative aspects of culture results can also be brought to bear on cost savings efforts. For example, a single colony growing from a specimen obtained from a sterile area should be fully worked up. However, light growth of a member of the *Enterobacteriaceae* family from a sputum specimen does not usually imply clinical significance and, therefore, does not require further processing.[7] Additionally, light growth of most microorganisms from non surgical wounds may not require further processing and identification.

Within the last few years, there has been a virtual explosion in clinical microbiology laboratory identification systems. The role of the new technologies has been to increase accuracy and to reduce the time required to identify or otherwise characterize microorganisms. However, the new systems have been unable to reduce significantly the cost to the patient for the processing of microbiologic specimens. Laboratories are generally expected to identify to species, and test for susceptibility most or all organisms present in a clinical specimen. It is ironic that proficiency testing programs such as College of American Pathologists measure the competency of the laboratory performance by the extent of complete identification.[20,30]

Although the greatest cost savings probably occur in the area of avoiding unnecessary work, the microbiologist should strive to improve the overall efficiency of the management of the workload. This can be achieved by the adoption of sensible, brief schemes for identification, and also through continued use of automation. Same day results may save administrative work in addition to having clinical impact. Some laboratories may find it advantageous to contract with other laboratories for identification or susceptibility services that they cannot offer efficiently such as mycobacteriology, mycology, specialized assays, and anaerobes. Larger laboratories may find that their staffing and automation will allow them to take on an increased volume of tests without corresponding increases in personnel. These tests are not likely to come from their own institution, but might come from the smaller laboratories that wish to cut their costs by arranging referral contracts.

To achieve the maximum effect, the emphasis on rapid testing must be related to clinical practice. Clearly, not all rapid results will lead to earlier decision-making on the clinicians part; for example, full identification of isolates available after 8:00 P.M. will not improve patient care since these results will not be acted upon differently. Finally, to have any rapid impact, the identification results must be communicated to the physician as soon as they are known. Establishing effective computer systems for the clinical microbiology laboratory should have the highest priority because they serve the laboratory, institution , and patient.

The following are general statements about the required extent of identification:

1 The extent of identification should be appropriate to the specimen type, the type of organisms recovered, and the needs of the physician.

2. The identification of species of staphylococci beyond *Staphylococcus aureus* is of little help to the physician, unless the isolate is from blood or other normally sterile sites.

3. The complete identification of *Enterobacteriaceae* from contaminated or non-sterile sites is often a needless expense and effort as identification of this organism may have little clinical impact.

4. Use spot tests or kit tests along with primary isolation plate colonial morphology and the spot indole test to presumptively identify isolates. The use of non kit and non-instrument rapid tests can greatly reduce the overall expense of the microbiology laboratory.

5. Definitive identification of all anaerobic isolates may not be required, and may not provide benefit to the physician. There are some situations where the presumptive identification of anaerobes is sufficient to allow diagnosis, proper therapy, and prediction of outcome; and these methods are reliable and inexpensive.[20,22,28] In other situations, however, complete identification is essential.

During the past decade new microorganisms have been identified as causes of infection, and old organisms have been found to cause new disease manifestations. New concepts have been developed for the identification and diagnosis of many infections; however, these procedures may be requested regardless of their cost effectiveness. Clearly, it is important for some laboratories to have the capability to screen for such unusual organisms; however, it may not be possible financially for all institutions to perform this service.

Examples of Cost Effective Workup of 11 Types of Clinical Specimens are Described Below.

Anaerobic cultures

Clinical microbiology laboratories should have the capability of culturing for anaerobes. Anaerobes are responsible for many serious infections and are resistant to many commonly used antimicrobics. Anaerobic isolates should be identified to the extent appropriate for the specimens source and the type of bacteria present. Limitations on anaerobic bacteriology procedures should be implemented based on patient care needs. The extent of anaerobe identification necessary by each clinical laboratory may be different depending on the needs of their physicians, the type of patient seen in their facility, and various operational and financial issues in the laboratory. The goal of the clinical microbiology laboratory, however, is to provide clinically useful information–and if the extent of anaerobe identification is decided upon only by financial issues, patient care may be compromised.

The cost of anaerobic cultures can be controlled through the development of protocols on what specimens to culture (See Table 3), how to transport the specimens, the development of rejection criteria on inappropriate clinical specimens, the use of good primary, selective and differential media, and the use of rapid presumptive identification techniques.[32]

Specimen quality is essential for good anaerobic bacteriology. Since many anaerobic infections arise from the existing normal anaerobic flora, the laboratory needs a specimen that is obtained without any contamination. This means that aspirates obtained by needle and syringe, or tissues obtained during surgery or biopsy will be of greater value than swab specimens which have a greater potential of being contaminated during collection. Certain specimen types are suitable for culture; in general, specimens with the potential of contamination with normal anaerobic flora are not suitable.

In addition, since many anaerobes are sensitive to the toxic effects of oxygen, proper transport in oxygen-free tubes is essential to prevent exposure to air during transport to the laboratory. Information is available[16] on a variety of transport systems, recommended transport times, and suggested volume of specimen. Once the specimen arrives in the laboratory, good anaerobic environmental conditions are necessary, such as anaerobic pouches, anaerobic jars, or chambers.[16] Furthermore, the use of good primary, selective, and differential media supplemented with vitamin K and hemin can actually save the laboratory labor and supply costs, and more importantly provide faster information to the physician.[16,32]

Presumptive or group identification may be all that is necessary on some anaerobes. The extent of identification should match the specimen source and organism recovered. For example, a specimen obtained from surgery, or from a normally sterile site should be identified as completely as possible so that this information may be used by the physician to understand treatment course, guide therapy, and understand the possible origin of the anaerobe. On the other hand, definitive identification of all anaerobic isolates may not be necessary since the patient may be currently treated with a broad spectrum antibiotic. Communication with the physician on mixed culture specimens may save the laboratory labor and supply costs.

Blood cultures

The culturing of blood for bacterial and fungal pathogens is among the most important functions performed by the clinical microbiology laboratory. Laboratories need to have the ability to provide the physician with prompt and accurate information on potential etiological agents.

Microbiologists need to control and provide measures to limit the number of blood cultures requested. Blood cultures from hospitalized patients should be limited based on clinical usefulness. In most patients with suspected bacteremia, three blood culture sets are sufficient to isolate the agent.[7,14,16] Ideally, these should be collected separately, at hourly intervals within a twenty-four hour period. Obtaining two sets of blood cultures within a twenty-four hour period has been shown to recover about 90% of bacterial pathogens recovered from blood.[7,14] On the other hand, a solitary blood culture is rarely useful for the diagnosis of bacteremia, and generally only wastes resources.

Blood cultures are one of the most expensive procedures in the clinical microbiology laboratory. However, rapid initiation of appropriate treatment based on blood culture results can save hospitals money and lead to more successful therapy. There is a difference of opinion about the most appropriate procedures to collect blood and choice of blood culture methods. The use of blood culture automation may provide positive results more rapidly to the physician by continuously monitoring the blood culture bottles, thereby detecting bacteremia hours before older conventional systems. The instrument is initially expensive; however, it allows more efficient use of laboratory personnel. The positive impact to patient care can and should be used to justify purchase of the instrument. The use of the lysis-centrifugation technique such as the Isolator (Wampole, Isostat Microbial System) is expensive and very labor intensive; however, it may have some usefulness in recovering very fastidious microorganisms, particularly mycobacteria and fungi in blood.

It is essential that blood for culture be collected aseptically. The cost of a contaminant to the laboratory and to the medical facility is significant.[1,14] Initially the skin at the venipuncture site should be cleansed with 75% alcohol followed by application of 2% iodine solution in a concentric fashion for at least 2 minutes. Adequate skin decontamination can reduce over-all hospital costs.

Pre-packaged skin preparation kits that contain the alcohol preps, the iodine, and a saturable skin brush (Medi-Flex, Overland Park, KS) are very effective. The initial cost of these preps seems high at 0.70 cents each, compared to about 0.15 - 0.20 cents for the alcohol and iodine separately, but the use as a kit with the brush has decreased our rate of contamination significantly. The cost in labor and unnecessary antibiotics for even one false positive patient per month is enough to easily justify the higher costs of the proper skin preparation kits. When a possible contaminant is recovered, it is best to report it out as "possible or probable contaminant" to alert the physician to draw new samples. The determination of whether or not an isolate is a contaminate is a clinical decision.[1,14] For example, if the skin site is not prepared properly, the laboratory may isolate coagulase negative *Staphylococci,* which in turn may prompt the physician to use vancomycin. The total cost, therefore, includes not only the laboratory cost (labor and supplies), but additional antibiotic usage and additional pharmacy charges, extended hospital stay, and additional hospital room charges. If false positive blood cultures are interpreted clinically as real positives, the cost to the health care system is staggering. One study by Bates[1] detailed the costs of 94 false positive episodes. This resulted in subsequent increased length of stay (median 12.5 versus 8 days) as well as increased pharmacy charges, and laboratory charges with an overall increased cost of $6,000 per patient to the hospital. As a consequence of contaminated blood cultures, the true cost of contaminants may greatly exceed those of the test itself.

The rate of contamination in a hospital setting may vary. For example, we consistently find that the rate of contamination for phlebotomy obtained blood cultures is significantly less than those obtained by the medical staff [1.0 % contamination rate by phlebotomists vs 5.0% for (nurses and physicians)].

The importance of sterile technique when obtaining blood cultures is essential to perform cost-efficient clinical microbiology, and these techniques require continuing in-service education. Data should be provided to the medical and other staff through newsletters and in-service instructions regarding the current rate of contamination, and suggested methods to improve this rate. This educational effort provides an excellent opportunity for the staff to be personally involved in improving the usefulness of blood cultures and controlling cost.

Body Fluid Cultures

Culture of normally sterile body fluid specimens should include inoculation of suitable media, and methods for the recovery of associated pathogens. The larger the volume of specimen cultured, the better the possibility for detection of the microorganism. As an adjunct to solid media, the inoculation of the body fluid into liquid media has improved recovery and the clinical usefulness of the test. The routine culturing of body fluids for *Legionella,* fungal, and mycobacterium sp. may be unnecessary. These organisms have a low prevalence of causing infection from body fluid specimen unless the physician has a high suspicion of these agents. The culturing of spinal fluid and pleural fluid for mycobacteria on a routine basis is impractical because the positive yield is usually less than 1%.

The use of a Cytospin (Cytospin, Shandon, Lipshaw, Pittsburg) is very helpful in making concentrated smears for Gram stains of specimens taken from body fluids. Although there is some up front cost associated with the instrument and some daily cost associated with supplies, the increased sensitivity of the Gram stain to guide antimicrobial therapy and reduce over-all hospital costs is justified.

Respiratory Specimen Cultures

The sputum culture undoubtedly is one of the most misleading of all specimens with regard to true clinical correlation. Specimens are frequently collected by personnel who may not be aware of the need for a fresh, uncontaminated specimen resulting from a deep cough. Consequently, many respiratory specimens are often contaminated with oral microorganisms and add cost to the laboratory and to the medical facility. On the other hand, the absence of pathogens does not exclude the presence of serious infection. For example, in only 50-60% of patients with bacteremic pneumococcal pneumonia can the organism be recovered from expectorated sputum, suggesting poor sensitivity of the respiratory culture.

It is essential to screen the sample. The most practical method is to screen the expectorated sputum by using the Gram stain to look for the presence of epithelial cells (which signify considerable oral contamination), the presence or absence of leukocytes and alveolar macrophages, and bronchial epithelial cells (which signify that the material was obtained from a deep cough from the lungs).[3,5,7]

A rapidly available Gram stain report can also direct the physician to use the most appropriate antibiotic. This may eliminate the need for more expensive broad-spectrum drugs and further testing. The receipt of one sputum sample per day will allow the diagnosis of most upper respiratory tract infections. If multiple or duplicate samples are received, and similar organisms are recovered from the multiple samples, the notation "see results of duplicate sample", or "see results on specimen___" should suffice instead of further workup.

Mycobacteriology Cultures

Tuberculosis still accounts for a large number of deaths and great morbidity worldwide. Disease from mycobacteria in the United States has steadily increased because of the acquired immune deficiency syndrome (AIDS) epidemic.

Laboratory personnel should provide guidelines and methods on the proper collection and transport of specimens for the recovery of mycobacteria. The selection of specimens for the detection of mycobacteria has evolved from respiratory specimens primarily, to stool specimens, blood, and urine samples, in addition to sputum or respiratory secretions in patients suspected of AIDS. Therefore, the laboratory must provide guidance on the type of holding medium, selection of transport containers, and the conditions under which specimens should be maintained.

Acid-fast bacilli (AFB) may infect almost any tissue or organ of the body. Pulmonary secretions are the primary sample and these include: sputa, induced sputum, gastric lavage, trans-tracheal aspiration, bronchoscopy, and bronchial lavage. Early morning urine samples are the specimens of choice for AFB. For the diagnosis of classic cases of pulmonary tuberculosis, at least three successive first-morning, deep-cough sputum samples should be collected. The patient, however, must be instructed on how to obtain this deep-cough specimen. Limiting the number of sputum specimens to once a day for three days still provides the clinician diagnostic value, and helps to keep the cost of AFB cultures under control.

The Bactec radiometric procedure for *M. tuberculosis* is now considered the method of choice as it provides rapid results and possibly increased sensitivity for the detection of mycobacteria. However, it is more expensive than the traditional agar method, although with some specimens such as blood in which the number of organisms arc small, the Bactec procedure is preferred. Other methods include the lysis-centrifugation technique (Isolator).

Properly obtained samples, adequate decontamination, and concentration procedures are essential. Detailed information on many of these procedures may be found in many textbooks on clinical microbiology.[7,16]

Mycology

The proper collection of specimens and their transport to the clinical laboratory are of major importance for the recovery of fungi, and for providing rapid, accurate results for the diagnosis and management of mycoses. Since overgrowth with contaminating organisms is common, it is important to transport the specimens to the clinical laboratory as soon as possible. Methods on the proper collection, transport, and reporting practices are available.[7,16]

The routine use of media for the isolation of fungi is not cost effective. It is best to perform fungal cultures when the test has been ordered by the physician or when knowledge of the patient's symptoms or history suggests a fungal infection.

The identification of the fungal isolates requires labor-intensive procedures, and reports are often delayed because of growth characteristics. These factors make mycology slow and expensive. On the other hand, fungal agents are becoming more prevalent in some immunocompromised patients.

Mycology results depend on quality specimens. The best way to significantly affect patient costs is to be sure that the specimen received is of good quality , that the information obtained from the specimen is rapidly reported to the physician, and that the report helps the physician to expedite proper patient care. Processing of poor specimens, production of questionable information from direct smears, and overly thorough identification of most cultures do not lead to cost-effective, high-quality patient care.

Parasitology

Fecal examination for the diagnosis of intestinal parasites (O&P) is one of the most expensive and labor-intensive tests in the microbiology laboratory, yet it is often one of the most abused requested tests in the clinical microbiology laboratory.

A very low incidence of positive results are obtained from the O&P test, and the majority of positives have been from outpatients.[20,24] It may be more useful and more cost-efficient to initially screen in-patient samples for *C. difficile* toxin then perform the O&P if the toxin test for *C. difficile* is negative.[25,29] *C. difficile* is by far the most common cause of microbial diarrhea, from 14 to 22% of cases, and unfortunately there are no good rejection criteria.

The reported yield on O&P samples on in-patients with more than 3 days of hospitalization approaches 0%, and the yield in patients with fewer than 3 days of hospitalization is about 2%.[21,25] The greatest yield is from out-patients which account for about 5% of positives.[21,25] Some investigators have reported that performing a *Giardia* antigen detection test may be the single most cost-effective procedure since *Giardia* is the more commonly recovered organism. Siegel et al.[29] noted that outpatients and patients who were in hospital less than three days did not require multiple stool examinations as had been the standard procedure. Only 9% of the patients required more than one O&P, and none required more than two specimens for diagnosis. If *Giardia* was suspected clinically, a second stool could be submitted no sooner than 5 days later. Patients with diarrhea caused by amoebae or flagellates generally have trophozites in the stool specimens. Reliable detection of trophozoites requires the use of permanent stains. If this is not done, the diagnosis may be missed. Shedding of parasites such as *Giardia* cysts, *Strongyloides* larvae, and schistosome eggs may vary from day to day. Therefore, it is recommended that up to three specimens be examined, with specimen collection every two to three days. Some investigators have also reported that with the exception of *Cryptosporidia*, no other pathogen was detected with the trichrome stain that was not also observed in the direct or concentrate sample.

Personnel is the most important factor in the performance of good parasitology. The laboratory diagnosis of parasite infection is predicated on the skills of the personnel. Many laboratories see little positive material. Personnel may have had marginal training. Results of the College of American Pathologists (CAP) Parasitology Survey program[30] show that mean values for performance of some parasite identifications, such as *Giardia lamblia,* have improved in recent years. However, for some other parasites there has been little change. If laboratories are not proficient in identification of parasites, it is misleading to offer this service. It might be better to fix specimens promptly in appropriate fixatives, and then forward them to a reference laboratory. Although this approach may not be the most economical, since the laboratory is spending money for the use of reference laboratories, the quality of results and improvements to patient care will increase.

There has been a changing spectrum of parasitic diseases in the United States in recent decades. *Strongyloidiasis* may be life-threatening to the growing number of immunocompromised patients. *Schistosomiasis* is the most important helminth infection world wide, and may represent a problem in immigrants and other foreigners in the United States as well as in U. S. citizens who have resided in or visited endemic areas, particularly if they have been exposed to surface water, as may occur in river rafting, swimming, or bathing. There is an increased incidence of protozoan diseases in the male homosexual population and in day-care centers. *Giardiasis* increasingly is recognized as a serious problem in the United States, particularly with respect to water supplies derived from surface water that is not filtered. Day-care centers are also a serious risk area because of poor hygiene in young children. *Cryptosporidiosis* is a problem, especially in AIDS patients, but there is increasing evidence that it causes diarrhea in healthy persons in an epidemiological pattern similar to that of *Giardiasis*. AIDS patients or patients with animal contact, such as veterinarians and farmers, will likely have greater yields of *Cryptosporidium*. However, the disease is present in the general population. The laboratory may wish to limit the examination to watery, frothy stools and highly susceptible patient populations.

It is speculated that in the future, parasitic detection will be performed by the use of pooled parasite antigen detection kits, particularly for persons with *amoebiasis* and *Giardiasis*. It may be possible to perform the protozoan portion of the examination by antigen detection using immunodiagnostic techniques. Specimens would need to be screened only for eggs, which are much more readily identified. Parasitic diagnosis is likely to remain expensive and of marginal quality unless these diagnostic activities are consolidated, or unless technology provides non morphologic methods for diagnosis.

Other cost-efficient methods for the detection of parasites may be to screen samples for *Giardia* by EIA, this is cheaper than performing the traditional O&P exam. EIA has been shown to be more sensitive than O&P, and it may be possible to combine and then screen several patient samples together. Then, if the pool is positive, each patient is tested separately. Communication with the physicians on the sensitivity of *Giardia* EIA, so that they can understand the limitations and significance of the test, is important.

Serology

Serological tests may be employed to detect the presence of an immunological response in the patient's serum. Antibodies can be detected through the implementation of various tests. The reader is directed to textbooks on clinical microbiology for further information.[7,16]

It is important when considering various serological tests to have knowledge of the sensitivity and specificity of the tests considered. Once the test has been understood, recommendations for the turn-around time and performance must be made available for the staff. Recommendations on proper performance must be determined prior to initiation of the test, and then the information must be fully communicated to the medical staff.

Depending upon the volume of the specific test request, it may be more economical to send some serological tests out to a good reference laboratory.

Stool Culture

The request for culturing of stool specimens for enteric pathogens (*Salmonella, Shigella*, and *Campylobacter)* on patients with gastroenteritis is often abused and misunderstood.[11,17,29.] Stool samples received for routine culture should be examined for the presence of *Campylobacter, Salmonella,* and *Shigella* sp. Only if the incidence of *Yersinia enterocolitica* gastroenteritis is high enough in your area, should this agent should be cultured routinely. Studies have shown that routine culture of stool specimens for *Yersinia enterocolitica* is not a cost-effective procedure.[17] For routine cultures, the use of an additional selective medium for detection of small numbers of gram-positive organisms (primarily *Staphylococcus aureus*) does not seem warranted. The isolation and identification procedures for *Camplylobacter* require specific media, incubation at 42°C, and microaerophilic conditions. The isolation of *Salmonella* and *Shigella* is carried out at the same time.

For the primary isolation of intestinal pathogenic *Enterobacteriaceae*, most laboratories use a minimum of two types media, differential media and high selective media. Some laboratories advocate the use of enrichment broth to assist in the isolation of *Salmonella* and *Shigella* from stool specimens containing a mixed bacterial flora. Use of enrichment broth is not indicated in the culture of *Shigella* or *Salmonella*; however, some laboratories persist in using these broths even though Shigella may not survive enrichment. Inclusion of an enrichment broth for the processing of stool specimens was once a requirement of the College of American Pathologists (CAP) and other inspection agencies, however, the recognition that enrichment broth was sometimes of debatable benefit led agencies to discontinue this policy. The current requirement is that laboratories should use enrichment or differential media.

Once positive cultures have been identified, an array of biochemical tests can be used to confirm the presence of *Salmonella* or *Shigella.* After the presumptive identification of the presence of *Salmonella* or *Shigella* has been made, the technologist must perform biochemical serological confirmation. Since the overwhelming majority of fecal samples sent to the clinical microbiology laboratory for testing are free of enteric pathogens, a screening test such as a latex test from Wellcolex with a high negative predictive value can eliminate further testing. Such a procedure can result in savings in labor and prevent subsequent testing. Multiple colonies may be pooled and screened in a single reaction. The cost is about $4.20 each for *Salmonella* and for *Shigella* screened.

The routine screening with special media of some enteric pathogens may not be appropriate, as in the case of *Aeromonas hydrophila.* The media used by most laboratories for stool culture will support the isolation of *Aeromonas*. If colonies are screened by the oxidase test, *Aeromonas, Plesiomonas,* and *Vibrio cholerae* may be discovered. *Yersinia enterocolitica* can be detected in many cases without special media.[17]

Some studies report that the presence of fecal leukocytes is not a useful predictor for the presence of enteric pathogens. Therefore, not only does initial screening of samples by Gram stain or other staining techniques fail to allow selective plating, such staining technique may also have little or any diagnostic value to the physician.

Proper specimen collection and delivery are essential for optimal recovery of enteric pathogens from stool specimens. Specimens that can be delivered to the laboratory within 1 hour may be collected in a clean, waxed, cardboard or plastic container. If a longer delay is expected, the specimen should be placed in transport medium. Cary-Blair transport medium has been shown to preserve the majority of stool pathogens the best. The yield on stool cultures performed on patients hospitalized longer than three days is extremely low.[11,29] Therefore, it is not recommended to culture these stool samples.

Wounds, Miscellaneous Specimens

Wound cultures, derived from lesions that are contaminated with intestinal, cutaneous, or oropharyngeal flora, result in unnecessary effort and produce reports that imply infection, and therefore suggest to the physician the use of antimicrobial therapy. It is not economically feasible to identify every organism present in wounds. This only leads to misleading information which can be misinterpreted. Interpretation and susceptibility testing must be based upon the quantity of organisms present and the specimen source. The proper collection of wound samples cannot be over emphasized and is essential for correct interpretation of results[5,7,16] Tissue or fluid obtained from a site presumed to infected is always superior to swab specimens, and in general, the more specimen the better. Communication between clinicians and laboratories is essential so that the physician knows the appropriate methods for specimen collection. It is best to perform both a smear and culture routinely on wound specimens. Most importantly, the laboratory must screen specimens by initial Gram stain smear for quality as well as to provide early results to assist the physician with the choice of antimicrobic therapy. Furthermore, quantitative or semiquantitative reporting of culture results is recommended. This provides some information as to the relative importance of various organisms present in a mixed infection.

Urine Cultures

Prevention of contamination by normal vaginal, perineal, and anterior urethral flora is an important consideration for collection of clinically relevant urine specimens. As many as 60% to 80% of all urine specimens received for culture may contain no etiological agents or may contain contaminants only.[8,35]

Procedures have been developed to identify quickly those urine specimens that will be "negative" on culture, thus circumventing excessive use of media and technologist time. The Gram stain is the easiest and most reliable screening method for identifying urine samples that contain greater than 10^5 CFU/ml. Several chemical tests can be used to assess parameters that may indicate urinary tract infection. Different systems have become available for the rapid detection of bacteriuria.[16] Some of these are as simple such as strip detection of nitrate reduction and leukocyte esterase, others require equipment or bioluminescence. On some of these tests the sensitivity is quite low. In general, there is a cost advantage to screening only in laboratories that receive many culture-negative specimens. Since most clinical laboratories receive more urine specimens than specimens of any other type, it is important that laboratories develop an efficient, cost-effective way of handling them. Some form of screening seems to be an important strategy for most laboratories.

However, the use of many of these rapid or screening urine tests have minimal impact on patient care, including out-patients for whom empiric therapy is usually successful. Rapid urinary detection systems , therefore, do not appear to have an over-whelmingly practical clinical impact on hospitalized patients. The clinically practical approach is to screen urine with inexpensive strip reagents, issue a preliminary report, and monitor the results of a conventional urine culture.

A scheme which incorporates the number of organisms isolated from the primary plates may be used to assess the potential significance of isolates, save time, and reduce expenses. For example, isolation of two or more species is considered most likely to be indicative of contamination, unless previous information about the patient suggests otherwise. In addition, the workup of urine cultures could be based upon the gross colony count. For example, a urine sample with two isolates in which one has a colony count of $>10^5$ could be identified with susceptibility testing, if the other isolate with a lower count such as 10^3 could be reported out only on the basis of gross morphology (lactose-fermenting gram negative rod, or coliform) instead of using more expensive procedures.

An economical workup on urine samples may be as follows:

•Urine screening. Whether the use of automated urine screening systems are accurate enough to replace the urine culture, and whether they have the ability to define and eliminate negative specimens rapidly needs to be determined. A urine screening device would be useful only if the time and cost to use it are comparable to or less than the conventional culture method. The direct cost of supplies for screening a urine sample culture is approximately 40 cents. The Chem-strip LN (Biodynamics, Indianapolis, Indiana) costs about the same; however, the sensitivity of these strips (in the <90% range at 10^5 colony-forming units/ml), is not high enough to screen urine at most hospitals and in most out-patient populations in which a high prevalence of symptomatic patients occurs. All of the other urine screening tests that do not require the acquisition of major capital equipment cost three to five times more than the present methods, and would create for this hospital an additional cost of $40,00 to $80,000 per year.[7,18,30]

•Use of spot tests and colonial morphology from primary isolation plates. See previous sections on the use of spot tests and colony morphology to provide cost-efficient presumptive workup.

• Use of non-kit and non-instrument rapid tests are also other methods to provide cost-efficient identification methodology. For example, screen Beta-hemolytic group A and *enterococci* with PYR; screen *Staphylococci* with the coagulase test.

Extent of Susceptibility Testing on Technical Operations

Susceptibility testing of rapidly growing bacteria is perhaps one of the most useful procedures that the microbiology laboratory can offer the clinicians and may be more important than definitive identification. In some cases, however, a substantial amount of the susceptibility testing that is performed in clinical laboratories is unnecessary and generates clinically irrelevant and potentially misleading information.[3,4,13,22,27,28] The decision to perform an antimicrobial susceptibility test belongs to the laboratory. A request should be thought as an approval rather that a signal to automatically perform testing on isolates when perhaps the culture is contaminated, the organisms isolated are not pathogens, and the results would not be clinically useful. In most cases, limiting the performance of susceptibility on up to three isolates make sense. When more than three isolates are found in equal numbers, cultures are usually reported as mixed.

Several methods are employed, from bacterial disk diffusion methods (Kirby-Bauer), manual minimal inhibitory concentrations (MICs) methods, to fully automated.[7,16] Automation has clearly made an impact on susceptiblity testing. The results are available to the laboratory worker quicker; however, the physician may not be available to receive the results, particularly at night. Rapid susceptibility testing using fluorescent markers and/or metabolic end products are available, although some rapid results may differ significantly from conventional standard methods. It may be very difficult to shorten incubation times to less than 6 hours and still obtain reliable results since some antimicrobial agent resistance, such as that caused by inducible B-lactamases, are not reliably detected with shorter incubation times.

Determining appropriate treatment for an infectious disease requires both the isolation of an infectious agent from a patient and the determination of susceptibility or resistance to antimicrobial agents used in therapy. Since the susceptibility of many organisms to antimicrobial agents cannot be predicted, testing individual pathogens against appropriate antimicrobial agents is often necessary. Clinical microbiologists recommend therapeutic agents based on their *in vitro* activity. It is often critical for the physician to accurately know whether the microbe being treated is susceptible or resistant to the chosen antibiotic.

Recent studies[2,27,28] estimate that 43% of all patients hospitalized in the United States received an antibiotic sometime during their hospital stay at a cost of up to $750. There are problems of overuse and misuse of these drugs. Accreditation and regulatory agencies in the past decade have required United States hospitals to conduct programs for improving the use of these agents. The Joint Commission on Accreditation of Hospitals (JCAHO) has been especially interested in this area, and has provided extensive and specific guidelines.[20] Under prospective reimbursement, the pharmacy, as well as the laboratory, have become cost centers rather than profit centers. This has led to increasing interest in ensuring that an antibiotic is used only when necessary, and only when equivalent, less costly drugs are unavailable.

Not all isolates need susceptibility testing, and in fact, reporting out some data sends a message from the laboratory that the organism is clinically significant when in fact it may not be. The microbiology laboratory needs to participate in programs to improve use of these drugs by working with the pharmacy and an infectious disease specialist. In a large hospital, an infectious diseases specialist and/or an active infection control committee and/or quality assurance committee will need to work together toward improving the utilization of antimicrobics, while the smaller hospitals or clinics may place more of the burden of this effort on the microbiologists.

Circumspect susceptibility testing can reduce hospital, pharmacy, and laboratory costs by preventing unnecessary antimicrobial therapy and directing appropriate and most cost-efficient therapy. The agents reported routinely for susceptibility testing must reflect the antimicrobial drugs available on the formulary. Costs can be reduced by restricting the reporting of results to pre-selected drugs. For example, the laboratory may test a member of the first, second, and third generation of cephalosporins, but report only the findings from the first generation agent if the bacterium is susceptible to that drug, or only report out if the organism is susceptible to gentamicin, instead of all aminoglycosides.

Role of the Laboratory in Improving Use of Antimicrobial Agents

The reporting of results of susceptibility tests in a clearly understood format is essential, however, the type of report that is appropriate may vary by institution. Minimal inhibitory concentration (MIC) data may be useful as a routine part of the report in hospitals with infectious disease specialists and others who make use of MIC data in their therapeutic deliberations.

Providing susceptibility summaries and antibiogram data to physicians is also essential.[22,27,35] Empiric therapy is the administration of an antimicrobial agent or group of agents prior to determining the identity or susceptibility of the organisms involved in an infection. This type of therapy constitutes a major part of hospital practice today. In one recent study,[22] over 80% of patients with infections were treated empirically. Success in empiric treatment depends in large measure on knowledge of the organisms likely to cause infection, and the susceptibility patterns currently encountered in these organisms. Timely summaries of infecting organisms by site and their susceptibilities can guide and improve therapy.

Optimally, antibiotic summary information should be provided to physicians through a laboratory computer reporting system, instead of printed information that may be out-of-date. Susceptibility patterns must be kept current, and costs and dosages should also be provided on the report. Decisions about what antibiotics to place on the formulary should be based on more than microbiologic data. Although such data is crucial to these decisions, other factors (cost, pharmacology and toxicity) should be considered as well. Microbiologists should provide data on susceptibility and assessments of the data which can help reduce the formulary. Sequential reporting of susceptibility data is important to reduce over-all costs. Some tested drugs are identified for first-choice use because of their pharmakinetics, or because they have a relatively lower potential for toxicity, lower cost, or less likelihood for producing drug resistance. If the organism is resistant to a first-line group of antibiotics, then a second group is tested and/or reported. A useful cost savings approach for the laboratory and the hospital[22] is to simultaneously test antibiotic groups, but sequentially report them based upon their resistance or absence of resistance.

Controversies with Susceptibility Testing

Accurate and reliable susceptibility testing of isolates is an important responsibility of the clinical microbiology laboratory. Laboratory services which can detect extended antibiotic resistance and the potential for treatment failure, can actually save the hospital considerable expense. Microbiology expenses in some aspects of susceptibility testing can save cost in patient days later on. Furthermore, it is important for the clinical microbiology laboratory to have a knowledge base of actual or potential organism resistance problems.

There are some organisms that are a problem for susceptibility testing due to either *in vitro* susceptibility techniques, or because of extended beta-lactamase production. For example, the susceptibility testing of *Enterococci* when recovered from non-sterile sites and from urine may not need to be performed in all cases, although with the frequency of recent vancomycin resistance among *Enterococci,* it may be wise to have available some approach to test clinically significant isolates. Susceptibility testing on *Enterococcus sp.* will be more clinically useful to the physician if the laboratory has the capability to enter a comment such as, "on serious infections the use of ampicillin or vancomycin plus aminoglycoside is strongly recommended."

The susceptibility testing of *Streptococcus pneumococci* may also be a problem. This organism was generally thought of as universally susceptible to most antimicrobics, particularly penicillin, however, resistance to commonly prescribed agents has been reported[7], prompting laboratories to perform tests on blood, CSF, and sterile sites. The use of automation to perform this testing has been unreliable to date; therefore, manual, expensive methods are employed.

The detection of extended beta-lactamase production with some automated susceptibility testing methods has been a problem with some "drug-bug" (antibiotic-organism) combinations. The use of automated systems can greatly improve the cost-efficiency of the laboratory; however, these systems must have the capability of looking for resistance as part of their computer-based expert software systems. The detection of these antibiotic resistance mechanisms needs to be reported and communicated to the physician for possible clinical impact.

The decision to perform susceptibility testing on anaerobic isolates is a controversial issue in the sense that there is no universally accepted "gold standard" against which to reliably compare test results. In addition, the recovery of some anaerobic organisms from mixed cultures may not provide meaningful data to assist in the decision of correct therapy. Therefore, the time and expense required to perform anaerobic susceptibility testing on some isolates may not improve patient care. There are situations, however, in which the microbiology laboratory needs either to be able to perform some level of susceptibility testing on clinically significant anaerobic isolates, or to be able to send these isolates to a reference laboratory. Specific infections from the following sites should be considered for susceptibility testing: bacteremia, endocarditis, osteomyelitis, joint infections, prosthetic device infections, and infections that do not respond to empiric therapy.[32]

The laboratory must also decide what antibiotics to report out initially. Of course, all pharmaceutical companies want their antibiotic to be reported, but on the other hand, it is important to report out only a limited list of the primary antibiotics commonly used in your institution. This limited list can direct the physician to the appropriate and most economical therapy. There is a significant cost to the medical facility when an inappropriate or incorrect therapy is used.

The Use of Automation for Susceptibility Testing

Many new systems for performing and reporting susceptibility test results are available to the laboratory. Much controversy exists over which systems offer the "best" results; i.e., which system is faster, which system is most accurate compared to standard techniques, and which system is most user-friendly.

Currently, a significant controversy exists regarding the problems associated with testing some non-fermenting oxidase-positive Gram-negative rods. Many results are unreliable. Furthermore, the automated susceptibility testing of *Streptococcus pneumoniae* apparently does not pick up resistance and therefore has some pitfalls for routine use.

Although rapid reporting is an important step in the successful management of infected patients, rapidly reported data is of little value unless the patient directly benefits.[27] It is essential, for example, to provide pharmacists a real-time computer link between the pharmacy and microbiology laboratory. This software is necessary to screen all patients automatically for mismatches between pathogens and drugs, or to screen for doses inappropriate to minimum inhibitory concentration levels, renal function tests, or for other clinical contraindications. Attention should be paid to identification of opportunities to target a more appropriate narrow-spectrum regime after culture results become available. The cost savings are significant. In one study, the combined efforts of clinical pharmacy, microbiology, and infectious disease personnel successfully optimized antimicrobial therapy on a hospital wide basis which lead to a decline in expenditures of >$200,000 per year.[27] Correct antibiotic utilization, which leads to improved patient outcome, provides a cost savings to the medical facility which covers the cost associated with the purchase of any hardware and software to perform this service.

On some automated susceptibility testing instruments, the intent is to shorten the length of incubation time to allow more rapid turn-around time to the physicians. However, it has been shown that the reliability of some data diminished with shorter incubation times. It may be very difficult to shorten incubation times to less than 6 hours and still obtain reliable results. Some antimicrobial agent resistance mechanisms, such as that caused by inducible beta-lactamases, are not reliably detected with shorter incubation times.

The following are specific guidelines for cost-effective susceptibility testing in the clinical laboratory.

- *Enterococcus sp.* Generally, your laboratory may not need to perform sensitivities on urine or from mixed specimens. Speciate and perform sensitivities on sterile sites.

- Modify reports as needed to provide information to physicians on extended resistance of organisms. Add comments, as needed; for example, for serious infections with *Enterococcus sp.* "the use of ampicillin or vancomycin plus an aminoglycoside is strongly recommended."

- Include antibiotics, such as some third-generation cephalosporins to determine underlying resistance; some beta-lactamase production may show extended spectrum and potential problems not detected by other methods.

- Use a computer system to find and track resistance.

- Assist the physician with susceptibility test results. Communication with physicians on susceptibility testing results is vital. Interpretation of results helps patient care.

- Understand antibiotic resistance throughout your facility by performing susceptibility summaries.

- Look for beta-lactamase production and perform susceptibility testing of *S. pneumococcus* from blood, CSF, and sterile sites.

- Selectively report antimicrobials within drug families and by specific drug-organism combinations.

- Use automatic computer-delivered susceptibilities to patient-care areas.

- Flag results that are unusual and could seriously affect patient care.

- Report cost information with susceptibility results.

- Report predictive sensitivities based on institutional antibiograms.

- Report specific drugs for therapy based on Gram stain, preliminary and final culture information.

Useful Data for Physicians

Innovative approaches are needed to improve cost-effective antibiotic use in hospitals. Initially, empiric therapy is selected based on clinical impression, underlying disease, and the Gram-stained direct smear of clinical specimens. Subsequent reports from the clinical microbiology lab and/or specific data on the susceptibility of isolates help to guide the physician with the specific therapeutic regimen. However, other data on reports which compare antibiotic cost, dosing frequency, and pharmakinetics may also be extremely useful to physicians.[6]

Guidelines for Susceptibility Testing

The cost to the hospital of dealing with resistant organisms and the danger of their spread to the community can be significant. Unless proper laboratory surveillance and reporting practices are in place in the laboratory, patients may develop infections with resistant organisms which lead to prolonged hospitalization, use of more expensive antibiotics, and possibly death. The clinical laboratory must monitor the use of antibiotics, note signs that resistant organisms may be developing, and take action to prevent the expense of spreading nosocomial infection to the community. Therefore, the cost of performing some aspects of susceptibility testing may be minor compared to the potential expense to the medical facility if this is not done.

Automation and Instrumentation

Automated instruments generally increase laboratory productivity and improve accuracy and turn-around time. Although microbiology is the least automated area in the clinical laboratory, automation in clinical microbiology has steadily increased over the last twenty years. The automated instruments available for the clinical microbiology laboratory allow rapid detection of the presence or absence of infectious agents, or have the capability to detect metabolic end-products. Not only are there instruments for antimicrobial susceptibility testing, but it has been estimated that currently about one half of the hospitals in the United States use some type of an automated method for processing blood cultures. Automation in microbiology has found a place in today's cost conscious environment, particularly in smaller laboratories where specimen volume is too low to justify sufficient numbers of specialized technologists who perform only microbiology related testing.

Although the technology is changing rapidly, a partial list of automated instruments available to the clinical microbiology laboratory includes growth detection systems, urine screening devices, microbial identification systems, susceptibility testing systems, high-performance liquid chromatography, ELISA techniques or other immunological detection systems for the presence of antigen or antibody, flow cytometry, and instrumentation to measure antimicrobial levels in body fluids.

The microbiology laboratory must decide 1) whether to automate a given procedure, and 2) whether to purchase or lease the necessary instrument. Published reviews by the American Society for Microbiology of automated systems are very helpful.

Justification for the use of automated instruments should be based on decreased turn-around time, improved service, improvements in patient care, sensitivity and specificity, and increase in productivity. You must consider carefully the true cost of the instrument, daily supply expense, QC, maintenance, and potential down-time. A cost analysis of the instruments may show that supply cost is increased with only a small decrease in the number of FTE's (full time employees) required to perform the automated test. For example, when Stanford University Hospital wanted to automate its blood culture procedure, we discovered that the FTE actually saved was between 0.2-0.3 FTE, which by itself could not offset the expense of the instrument. However, there are other factors to consider. One justification is that many automated instruments may be interfaced directly to the hospital or laboratory computer, allowing the physicians to access the information faster. Also, the frequency with which the bottles are examined may promote faster detection which can save the hospital expense by decreasing patient stay.

The sensitivity, specificity, and predictive value of any new methodology must be completely understood before the procedure is brought into the laboratory to replace other techniques. For example, although new molecular technologies have the potential to enhance the speed, sensitivity, and specificity of microbial detection, these methods will tend to increase rather than decrease costs. New technology, because it is generally expensive, is a major cost containment problem, and initial enthusiasm must be weighed carefully against actual cost and usefulness of the new method. Although new technology may increase productivity, it is essential to determine the clinical usefulness of the information yielded by the new method and the accuracy and interpretation of the results. In addition, it is frequently difficult to evaluate the value of new technology when it first becomes available. Ideally, new technology should be implemented only when its benefits to the physicians and patients outweigh its cost.

The Use of Laboratory Computer Information Systems

Computers are a very cost-effective tool with many applications in the clinical microbiology laboratory that directly improve patient care.

Probably the greatest use of computers in clinical microbiology is in the prevention of errors or mistakes during input of data or susceptibility results. The computer can flag unusual results, which allows laboratory personnel an opportunity to recheck the results and catch mistakes before they are transmitted to the medical staff.

The second most important use of computers in the clinical microbiology laboratory is the reduction of the number of specimens and the elimination of duplicate specimens, both of which are essential to reduce laboratory costs. For example, computers can be used to notify medical personnel if a stool culture is ordered for a patient who has been in the hospital more than 3 days. Also, a computer can flag a second specimen on a single day, and thereby alert the laboratory to a potentially unnecessary test request. Furthermore, computers allow the laboratory to provide instructive comments on the acceptability of specimens. Comments such as "consultation with the laboratory is required before this specimen can be processed" or "duplicate specimen received" can help educate the physician as well as reduce laboratory cost. Such comments will verify the clinical microbiology laboratory's clear statement of policy which can also be computerized and transmitted to the medical staff to prevent any misunderstandings about test request policies.

Another advantage of computers is a decrease in the time required to make information available to clinicians. This is particularly the case when remote terminals allow immediate access from the medical units. Such terminals allow rapid, direct test requesting from the hospital units. In response, the computer can provide the correct specimen request form and ordering information which give detailed guidelines on specimen collection and handling procedures, and even produce a historical review of a patient's previous reports.

Direct communication between microbiology instruments and laboratory computers can allow data to be transmitted easily after verification. For example, computers can generate cumulative susceptibility data and produce algorithms that can be used in guiding the work-up of specimens.

Finally, cost-effective measures that can be aided by computerization include the following: improved inventory of supplies; cost determinations, utilization trends, and productivity measurements for various laboratory tests; cost-analysis of procedures, including labor and supplies; susceptibility trends (antibiograms) and costs; and surveillance and epidemiology data for infection control practitioners.

Alternatives to Conventional Culture Techniques

In the past few years there has been increasing development of rapid methods for detection of microorganisms by using alternatives to conventional culture techniques. Most of these techniques rely on the detection of either chemical by-products or of antigens produced by the pathogen. The challenge to the clinical laboratory is to decide whether these tests have the sensitivity or specificity to be used routinely, and whether their use is cost-effective and clinically relevant. Such decisions are a most formidable task for the laboratory director, and should be made in consultation with clinicians to better define the manner in which these tests are likely to be used.

Antigen tests are rapid, but sometimes less sensitive than conventional culture techniques. For example, until antigen tests for all agents of meningitis are available, and the data exists to demonstrate that they are sensitive and specific enough to replace culture techniques, the expenditure of resources to perform these tests on CSF, in addition to culture, is probably not justified. On the other hand, latex agglutination tests for beta-hemolytic Streptococcus group A, cryptococci, rotovirus, herpesvirus, and the detection of legionella antigen in urine seem to be good use of technology. The detection of soluble antigens of *Legionella pneumophila* in a patient's urine, for instance, appears to be both sensitive and specific, and may be used to make the diagnosis of legionellosis days before the organisms can be cultured and often weeks before serological confirmation of the disease can be accomplished. Finally, some rapid techniques are sensitive enough to detect even non-viable organisms, such as DNA hybridization detection of cytomegalovirus.

It seems likely that more of the these rapid alternative techniques will be used in the hospital laboratory. However, in this era of necessary cost-containment, any new rapid techniques must be cost-efficient. The information provided must be accurate, sensitive, specific, and reproducible. Above all, the information must have clinical relevance.

The Clinical Microbiology Laboratory in the Big Picture

Cost-efficient clinical laboratory practice may also mean adding selected tests to improve clinical usefulness. For example, the cost of performing susceptibility testing on Gram-positive organisms is minor compared to the cost and care impact that are required to manage a patient with antibiotic resistance, such as vancomycin resistance. Being concerned with saving a few thousand dollars per year in laboratory costs, when the entire annual savings from not doing that test can be lost in a one or two day extended patient stay in the ICU if resistance is not detected, does not make overall economic sense. The value of this perspective is not widely shared nor understood by all hospital administrators. At times, administrators impose cost-reduction pressures on the clinical microbiology laboratory without fully realizing that some laboratory procedures are necessary to reduce overall hospital cost. Furthermore, the impact of correct laboratory utilization (numbers and types of tests ordered) needs to be examined as part of the overall cost-reduction efforts of the medical facility. Clinical microbiology laboratory procedures should not be thought of as a stand-alone expense, but should be viewed as a response to procedures practiced throughout the hospital.

A short range viewpoint can have an adverse impact on patient care. Clinical microbiology laboratories must extend cost analyses to length of stay and patient impact, and not consider just costs within the lab alone. The laboratory can spend an extra $5 for another plate or a rapid antibiotic susceptibility test and save $500 down the line. Laboratories must think system-wide, not just by selected units. For example, laboratory computers are invaluable to report susceptibility tests as soon as possible. This can spare the use of expensive broad-spectrum antibiotics and multiple-drug regimens, and may thus reduce hospital days. Laboratory computers should be set up to report automatically only the cheapest effective antibiotics for a given specimen.

The Use of Reference Laboratories

A laboratory must decide which tests to perform in-house and which methods to use, as well as which tests to send to an outside reference laboratory and what reference laboratory to use.

Tests requested by clinicians that are not within the routine procedures performed by the microbiology laboratory should be submitted to reference laboratories. The infrequent performance of a particular test is usually not cost effective, and may lead to inaccurate results because the skills required for their performance are not exercised with appropriate frequency. The nature and the number of tests submitted to reference laboratories should be periodically reviewed to determine if there are increases in volume that would make it more cost effective to perform these tests in-house. Sending tests to reference laboratories is not an admission of a poor quality laboratory and/or one lacking in microbiology expertise. Making appropriate use of a reference laboratory is a cost-effective practice.

Laboratories can remove low-volume tests from their menu to realize cost savings. Cost accounting information can help determine which tests should be sent out based on direct cost alone. Other costs to be considered are the cost of quality control testing, the supplies that must be discarded because of outdating, and costs of materials, proficiency testing surveys, and clerical time. These costs should be compared to the cost of having the test performed in the reference laboratory, as well as the cost of mailing or other transportation expenses needed for the delivery of the specimen. Non-cost considerations include the ability of an outside laboratory to provide results in time to be clinically useful, and certainly, the quality of the reference laboratory.

In contrast, there are some negative components which must be considered. At times, there is loss of prestige among in-house staff who no longer perform a test, and a change of prestige of the in-house laboratory among clinicians users and administrators who may use the information in public relation situations.

LABORATORY MANAGEMENT OPERATIONS

So far we have discussed patient testing factors (utilization issues, what tests should be ordered, and how and why to change ordering practices), and technical operation factors (the quality, collection, storage, transport, and processing of specimens).

Let us now focus on factors dealing with the management of daily operations in the clinical microbiology laboratory. What resources does the laboratory have to control and monitor its own cost to improve efficiency? What management ideas or techniques will yield the greatest cost efficiency in the operation of the clinical microbiology laboratory?

Look at all aspects of laboratory operation for opportunities to reduce costs. To begin with, be sure that the clinical microbiology laboratory is providing rapid, cost-effective results and reporting them promptly. The emphasis should be on rapid reporting, not just rapid testing. Next, check the staffing patterns. Check the equipment profile: is the laboratory as automated as it could be to produce medically necessary information and eliminate unnecessary expense for supplies and personnel? Consider such items as the test mix —the "make versus buy" decision. Finally, take into account the long term savings. Would a change now produce a beneficial result later?

Six areas of importance are discussed in this section: 1) analysis of operations, 2) measuring productivity, 3) negotiating contracts, 4) bench marking, 5) educational strategies, and 6) quality assurance and quality control. An understanding of these laboratory management operations will provide additional approaches for cost reduction in the laboratory.

Analysis of Laboratory Operations

Laboratory managers must have a basic understanding of cost-accounting, and must know the true cost of each laboratory procedure in order to monitor and control laboratory costs and performance. Almost all direct costs in microbiology laboratories are within two areas: 1) salaries and benefits, about 60-65% of the clinical microbiology budget, and 2) supplies, about 20-25% of the budget.

Microbiology, because it is labor intensive and not yet as easily automated compared to other areas of the laboratory, always incurs high labor costs. From a microbiology laboratory perspective, there is a direct relationship between work load and the number of technologists needed to perform the tests. The number of technologists may vary, depending on the presence or absence of automation in the clinical microbiology laboratory. Any reduction in the number of tests to be performed or in the time and effort required to perform a microbiology test will result in saved labor--and saved dollars. Therefore, the cost-efficient microbiology laboratory manager will carefully consider any rationale that leads to a savings in number of tests ordered, and labor and supplies required for a test.

At times, laboratory managers must rethink long-established hospital policies. Often some practice, seemingly small in importance, can compound to increase laboratory costs dramatically. For example, repetitive daily cultures from a suspected site of infection rarely contribute useful information for patient management. Excessive numbers of blood cultures, daily sputum cultures from patients without signs of pneumonia, daily cultures of various drainages, or cultures of superficial patient material are other examples of over-utilization of the microbiology laboratory's services. Although multiple factors contribute to excessive use of laboratory tests, physicians are the single most important factor in the expansion of laboratory test utilization. Education of the physicians and staff, therefore, may be one of the most important, effective cost-control mechanisms which the clinical microbiology laboratory can implement.

Some major factors to consider to improve the labor and supply costs in a clinical microbiology laboratory are as follows:

a) Examine fundamental laboratory operations. Decide what tests your laboratory should perform. Decide what organisms to identify and report out. Decide what media and reagents to use, and how and to what extent susceptibility testing is to be performed. Form consortia with other laboratories to buy items in bulk. Eliminate needless biochemical tests. Decide what media to use. Limit the extent of organism identification, which may by itself save cost. In addition, report out only clinically significant organisms, and report only brief taxonomic information on others (e.g., rare Gram-positive cocci), so that physicians do not react to clinically misleading information.

Careful analysis of laboratory practices can usually uncover some unnecessarily costly methods that can be eliminated or modified without compromising patient care. Often, laboratories become married to certain tests or procedures and do not want to change. In this new era of cost-containment and prospective payments, if the laboratory does not choose to change, administrators may step in and dictate a change.

Another important step is the calculation of accurate direct cost figures for every laboratory procedure. Monitor laboratory expense against budget allocation, and provide a baseline against which the cost of proposed procedural changes and new tests can be assessed.

b) Use cost accounting to find the best test methods. It is essential to know the actual cost (labor and supply cost) incurred in performance of each procedure. Monitor these factors periodically. All laboratories find themselves in a competitive and changing environment which is forcing them to become more selective and specific about test costs for both special and routine use. It is essential in this economic environment that laboratory managers know how to determine actual costs and estimate standard costs for tests based on their own equipment and methods. Cost accounting is important for laboratory management to identify costs for budget and control. Hospitals are being asked to look at many options including sending laboratory tests to reference laboratories to reduce costs. Before laboratory management makes a decision to "make or buy" laboratory tests, it is essential to know what it costs to do a test.

c) Re-think common laboratory practices, procedures and costs—things your staff does by rote; tests you seem married to. Your staff probably has many cost cutting ideas they would like to suggest and implement. In truth, the staff may know more about expensive, labor-intensive procedures that should be changed than the laboratory manager. Consider changes in vendors of supplies, service contracts, and which tests are sent to reference laboratories. Eliminate redundant proficiency testing and superfluous quality assurances practices.

d) Consider rearranging the work-flow or assignments. Reassign technologists from tasks that could be completed by a clerk, laboratory assistant, or secretary. Move work areas and equipment to maximize efficiency.

e) Consider changing your hours of operation. Possibly your laboratory could be more efficient if you changed the time when your staff started and finished. Opening a half of hour earlier might make your lab more efficient by spreading the work load more evenly. Likewise, changing the hours your staff stays in the evening may help get reports out faster, provide better service, and improve efficiency.

f) Examine your skill mix. Although some state regulations limit what responsibilities your staff may undertake, there are many tasks that can be shifted to employees with a lower base salary. For example, hiring more laboratory assistants may lower the overall payroll of your laboratory, but not change service, quality, or performance.

g) Use automation when possible. Earlier sections point out that automation may allow your laboratory to lower the overall payroll. Automation may improve the productivity of the laboratory because reports are available sooner to the medical staff. The quality of the results may improve as well.

h) Work with the medical staff to establish clinical pathways with an eye to eliminating unnecessary, superfluous tests, while maintaining clinical relevancy. Clinical pathways improve the utilization and selection of clinical specimens, they prevent the needless workup of duplicate samples and samples of questionable significance, and they decrease the volume of identification and susceptibility tests that have limited value to patient care.

Productivity Measurement

Another important step is calculation of accurate direct cost figures for every laboratory procedure, which allows monitoring laboratory expenses against budget, and provides a baseline against which the cost of proposed procedural changes and new tests can be assessed.

Cost Accounting in the Clinical Laboratories by the NCCLS[23], other information produced by College of American Pathologists Workload Recording Methods, and the newer CAP publication *Laboratory Management Index Program* (LMIP) are tools which incorporate workload recording and develop new productivity measures for individual laboratory sections and the entire laboratory. The NCCLS reference,[23] provides methods and terms to improve control of costs and to report significant deviations from standard cost.

Each laboratory should evaluate its information needs and the costs of acquiring and maintaining the information before beginning a cost accounting operation. Each laboratory should achieve a balance or level of comfort between cost and benefit, and simplicity and complexity of its accounting system to insure profitability and/or efficiency.

Other suggestions in *The CLMA Guide to Managing a Clinical Laboratory* [20], allow laboratories to look at the productivity, cost-effectiveness, and physician utilization of their operations, compared with laboratories of the same type (microbiology, chemistry, hematology, or transfusion medicine) and with the same number of output units. These methods provide cost data that help the laboratory manager to understand the relation of costs to individual test procedures. These data also enable laboratories to make informed decisions by providing a better understanding of the multiplicity of costs related to equipment, labor, and consumables. Careful analysis of laboratory practices can usually uncover some unnecessarily costly methods that can be eliminated or modified without compromising patient care.

There are a number of other factors to consider when evaluating laboratory operations for efficiency, such as service demands, culture and patient mix, volume, available instrumentation, orientation to rapid results, and hours of operation. Microbiology laboratories may need to rearrange work flow, expand the hours of operation, hire more laboratory assistants, and utilize more automated methods.

The use of the College of American Pathologists' (CAP) publications provide a resource for workload measurement. Using the CAP workload systems, it has been determined that in general, the small and medium-sized laboratories tend to process more specimens per worker than the large laboratories. This may be the result of the greater number of CAP workload units expended on specimens in large laboratories and possibly a larger portion of non-specified workload (non-productive time) in the large laboratories. In one study, the number of specimens per worker per year showed tremendous variability, regardless of the size of the laboratory.[26] The wide variation may be attributed either to under or over staffing, to the extent of specimen work-up, to the variability of workup by different technologists, or to patient mix or different service expectations.

Approximately twice as many specimens can be processed per worker in bacteriology and serology than in mycobacteriology, mycology, parasitology, and the virology sections of the laboratory. In general, the small laboratories expended less labor per specimen than mid-sized laboratories, which in turn expended less labor per specimen than large laboratories. The additional workload can be attributed to either more complex cultures, greater extent of available services, or lack of cost containment in the larger facilities.

In general, the number of specimens processed per worker has dramatically increased during the past decade, except in the large laboratories. The decrease in the average number of CAP workload units expended per specimen contributed to the apparent increase in productivity of the laboratory workers. The percent of time spent on non-workload activities has been dramatically reduced over the last few years.[20]

The use of clinical pathways in hospitals and ad hoc practice guideline processes that involve collective decision-making and consensus among laboratorians, clinicians, and nurses result in a tremendous reduction in utilization, as evidenced by a reduction of tests and more intelligent use of tests. Clinical pathways and physician-led multidisciplinary teams determine a standardized set of decision points and treatment protocols that provide the optimal quality outcome, which also reduce cost.

Laboratory managers need accurate reports on personnel costs, supplies, services, and overhead charged to laboratory. Almost all directs costs in microbiology as due to salaries, benefits and supplies, and these areas are the most appropriate to consider for cost containment. Workload units are important for assessing trends, rearranging staffing patterns, and forecasting future developments. Costs include salaries, fringe benefits, supplies, outside services, depreciation of capital equipment or renovations, interest, and overhead.

Once true costs, for clinical microbiology tests (costs/test) are available, the laboratory can then determine the break-even point for any particular test. This break-even analysis is helpful in making decisions regarding the volume needed for a test to be cost effective, whether to adopt a new methodology, whether automated equipment will enhance productivity or decrease the cost of the test, or whether the test should be referred to another laboratory. Cost-analysis of test procedures can have some additional value because third-party payers may base future reimbursement on costs rather than charges.

Contract Negotiation Techniques to Reduce Laboratory Costs

By understanding the tools of negotiation, the laboratory manager is in a better position to obtain the best contract price for supplies and reagents. Various services can help the laboratory manager determine the best contract price attainable. One such service is M.D. Buyline (M. D. Buyline, Dallas, TX), which provides pricing of supplies, equipment, and other consumables. The service sends information to the subscriber, or the information may be directly available by computer interface, as to the best price, along with various negotiation tactics, keys to successful negotiations, vendors' strategies, and other useful information to have prior to sitting down with vendors to discuss price. Laboratorians need to remember that salespeople are trained in negotiation tactics, but most laboratorians are not.

Four Key Areas for Successful Negotiation

1. Operate as a team– plan and then communicate rules to vendors
2. Keep the element of competition alive–either real or perceived.

3. Knowledge is power–Give as little information as possible; get as much as possible.
4. Keep a winning attitude–be confident, take the offense.

Other specific negotiation tactics that can be used with vendors to be cost effective are: the power of legitimacy (you have the authorization power and the right to obtain the best price, so use it), manage concessions (you have the right to control who yields, who gives in), control deadline (you need to set and control the contract deadline) identify weaknesses, the power of the "the flinch" (sometimes over-reacting makes a statement), the power of silence, split the difference, be willing to talk, get it in writing (all contract negotiations should always be in writing) including the consequences if the contract is broken.

Bench Marking

Bench marking is a technique that allows comparison of the financial and operational performance of many similar hospitals with yours. There are companies which provide this service (Mecon Associates, San Ramon, CA), or managers can directly call similar institutions and talk with the Medical Director to obtain data on financial and operational activities. This allows laboratory directors to identify "best practices" and areas of opportunity to improve, at least in comparison to other similar institutions. Direct comparisons of patterns of care, cost, services, FTE's, supply cost, and other financial operations can be made. Bench marking the relative financial and operational performance of hospital laboratories allows health care planners to design the most appropriate laboratory services system for future hospital inpatient and outpatient market demands. Factors to be compared are utilization, testing costs, staffing mix, productivity, and organizational structure.

Examining the financial and operational performance of similar laboratories should be thought of as a starting point. Once the data is compared and analyzed, it is also imperative to contact directly those laboratories in which performance appears best. Not only does this allow you to verify the data, it allows you to ascertain what differences, practices, or services exist in their institution that may account for their increased productivity. Furthermore, it allows you the opportunity to obtain their input on what changes in their operations may have worked the best to lower their cost.

Educational Strategies

Physician and staff education is important to improve the utilization of the laboratory. The services of the clinical microbiology laboratory, hours of operation, expected turn-around times, and specimen collection and handling procedures should be presented to new physicians and physicians-in-training at in-service training sessions. It is best if this is done directly by the laboratory staff, so that the physicians will know with whom they may interact and whom to contact for questions. This direct interaction and communication are important to create an atmosphere of teamwork.

A number of educational strategies may be developed to reduce inappropriate use of ancillary services and to improve patient care. Some may be successful, others ineffective. A positive physician attitude towards cost containment and an appreciation of the role of the laboratory in patient care is crucial. It is important to point out to them that improvement in the quality of care will result from cost control, because the concept of cost containment *per se* cannot be used as the sole reason for a cost containment program. Formal in-service instruction, dissemination of cost information, written guidelines or protocols on specimen collection, and feed-back audits of laboratory usage are useful. It is also important that physicians clearly understand how to properly request each laboratory test and supply what specific information the laboratory needs. If the physician does not do this, additional communication between the physician and laboratory will be needed and this adds to the costs.

In addition, a laboratory newsletter or on-line publication that includes discussion of timely issues, unusual susceptibility patterns, and current cost data can improve patient care and can contribute to cost savings.

Microbiologists need to actively and continuously educate physicians in the proper use of laboratory, the extent of diagnostic services provided, and expected turnaround time for test results. Furthermore, continuous interaction between the microbiologist and clinicians is essential to ensure that planned changes such as selection or deletion services, the reduction of identification methods, or the alteration of reporting of antibiotics do not have undue clinical implications.

Also, direct and continuous dialogue between the microbiology laboratory and the nursing staff is essential on proper laboratory utilization. The availability of detailed printed instructions on hours of operation, specimen collection and handling procedures, recommended transport containers, and number of specimens to submit can improve the quality of specimen submission and thus reduce cost. Also, a cumulative summary of antimicrobial susceptibility on microorganisms isolated from patients at the hospital should be available to the staff.

Although the physician is ultimately responsible for patient care, changing physician's test ordering practices, and improving specimen quality will reduce laboratory costs. Continuous effort is required to control incoming work load, but many laboratories do not have the personnel resources needed to implement and maintain a work load control mechanism. In addition, many laboratorians do not want to encounter the negative feedback received from physicians when recommendations are made to limit the specimens and to reject poor quality of specimens. Physicians may assume that all tests they request should be processed, and that all tests they request are clinically necessary, however, there are excellent studies to show that this may not always be the case.[3,5,10,13,19,31,35]

Cost of Quality Control and Quality Assurance

Clinical microbiology laboratories are regulated by a variety of agencies to ensure that testing is accurately performed. All of the steps beginning with initiation of orders for testing, collection and transport of specimens, and finally the timely reporting, interpretation , and use of the data are monitored to optimize the quality of patient care. Quality control and quality assurance regulations have been established and adopted from Healthcare Financing Administration (HCFA), the College of American Pathologists (CAP), and the Joint Commission on Accreditation of Healthcare Organizations (JCAHO), as well as from some State or County regulatory bodies.

Internal quality control in the laboratory helps to ensure that data generated by laboratory personnel are accurate, reproducible, and reliable. The performance of these activities involves continuing education, proficiency testing, the use of procedure manuals, and the maintenance of records on the performance of equipment, media, and reagents.[24,30] Quality assurance measurements look at how the data is being utilized by the medical staff and the impact on patient care.

Although, quality control and quality assurance are an essential part of the laboratory's work, these activities within the laboratory can affect the bottom line. The cost of quality control activity should be assessed and well understood. The clinical microbiology laboratory may have instituted excessive procedures or methods that have no impact on patient care. The laboratory needs to critically examine various surveillance procedures, and reduce the frequency and extent of these procedures when there are few or no deficiencies, while still meeting the regulations and standards of their regulatory agencies. For example, cost-effectiveness issues would include examining the frequency of testing media to be sure that NCCLS recommendations on testing frequency are met, testing only recommended media, and checking out new lots of selected dehydrated media and fresh batches. Other cost-effective measures would include examining testing frequency of biologic reagents, checking the performance of equipment, and monitoring temperature and susceptibility testing. The expense incurred by quality control in the clinical laboratory is justified only when it does not exceed the cost of an incorrect diagnostics. The cost of quality control has been reported as high as 4% of total bacteriology laboratory budget.[26]

The National Committee for Clinical Laboratory Standards (NCCLS M22-A) has placed a significant amount of the responsibility of quality control on the manufacturer . Many of the recommendations for quality control of media, reagents, and susceptibility have changed recently as a result of investigators questioning the cost-effectiveness of certain procedures. This has led to some modifications in the federal regulations.

Quality assurance programs, such as recommended by the Joint Commission on Accreditation of Healthcare Organizations (JCAHO), may be costly in labor and supplies. A clinical laboratory and hospital must continuously improve the clinical and support services that most affect patient outcomes. A program initiated by CAP Q-Probes, can provide the necessary support to perform intra- and inter-department review of practices and services related to the improvement of service. Another great feature is that this program provides information from all users throughout the United States. The initiation of a program such as this can provide the tools to improve patient care at a relatively low cost to the laboratory.

IV. CONCLUSIONS AND SUMMARY

The need for cost-containment and cost-reduction in the clinical laboratory will only increase. The misuse of laboratory services will not only be a cost-containment problem to the overall cost of health care, but will impact the way the laboratory is viewed by administrators and insurance companies. Insurance companies or HMO's will make more and more of the rules for test ordering and pricing which they will reimburse. This in turn will alter the extent of services provided by the clinical laboratory. Tests that are ordered without specific indications, or tests requested that not part of the insurance company's protocol will not be paid. It will be critical in the future for the physician to indicate clearly the clinical indications and need for tests. The laboratory that does not require this information will be increasingly at risk of losing revenue, and thus not able to survive financially.

The laboratory should initially examine and review the specimen collection and handling procedures. The microbiological workup of poor quality specimens only adds to the overall cost to the laboratory and to the medical facility as a whole. Furthermore, examining the extent of processing, testing and susceptibility testing practices will help to lower labor cost, although the limitations of test methodology and susceptibility practices are dependent on physicians needs. It is essential also to provide rapid reporting of results to physicians which may help overall laboratory and hospital cost savings. The aim of these practices would be to lower the labor involved with each test, but still provide clinically significant information, within the shortest period of time. Group buying and supply consortia will be mandatory to reduce the laboratories' supply cost.

Laboratory management operations will need to be reviewed to ensure that there is proper utilization of resources (labor and supplies). There should also be in place some system that allows monitoring of productivity and for comparison with similar institutions.

The greatest cost-saving will be in the area of utilization, and improvement of specimen quality. These two areas that should be modified to provide meaningful information without compromising service or quality. Controlling cost is more a complex issue and more integrated problem than simply asking the laboratory to watch the bottom line. Health care institutions are recognizing that you have to focus on the overall cost of patient care, not just the impact of clinical laboratory costs.

The true role of the clinical microbiology laboratory is to provide data to improve patient care, help control nosocomial infections and reduce length of stay. To accomplish these goals, the clinical microbiology lab must provide accurate data to physicians, monitor spread of disease, and provide timely and rapid information to the physician. The cost to the hospital of performing these functions is far less than the cost consequences of not performing them.

The concept, that clinicians submit specimens and laboratories run the requested tests without question, no longer holds true in this era of "cost accountable-patient outcome clinical laboratory testing". Specimen quality must be evaluated prior to plating and subsequent workup. In addition, patient parameters and other test results may indicate that certain test requests are inappropriate or incomplete, e.g., requesting latex antigen tests on CSF when the specimen contains normal levels of protein, glucose, and cells, or adding additional tests that the physician has not ordered. The clinical laboratory must now evaluate both the patient and the specimen to determine the most "cost accountable, clinically relevant " specimen procedures and decide which specimens and tests will be performed and which ones will be rejected.

Responsibilities of the Clinical Microbiology Laboratory to Reduce Cost

- Provide proper and appropriate procurement instructions, and ensure proper transportation of clinical specimens.
- Eliminate the misuse and overuse of laboratory testing.
- Ensure the quality of clinical specimens by initially evaluating if the specimen is appropriate and acceptable for culture and further workup.
- Institute rejection criteria and processing limitations.
- Institute methods for the rapid and cost-effective isolation, identification, and susceptibility testing of relevant clinical isolates.
- Provide laboratory services in relation to the "big picture" to reduce length of hospital stay, and to improve quality of care and patient outcome.

V. REFERENCES

1. Bates, D. W., L. Goldman, and T. H. Lec. 1991. Contaminant Blood Cultures and Resource Utilization. JAMA 265: 365-369.

2. Bales, M. J. McLaws, S. M., Feen, J., and Matsen J. M. Use of and cost savings with morphologic criteria and the spot indole test as a routine means of identification of *Escherichia coli.* Diagn. Microbiol. Infect. Dis. 1984; 2:187-191.

3. Bartlett, R. C. Medical Microbiology. 1974. Quality, Cost, and Clinical Relevance. John Wiley & Sons, New York.

4. Bartlett, R. C. 1975. Control of cost and medical relevance in clinical microbiology,. Am. J. Clin. Patrol. 64: 518-524.

5. Bartlett, R. C., and C. Rutz, 1980. Processing control and cost in Bacteriology. Am. J. Clin. Pathol. 74: 287-296,

6. Bartlett, R. C. Quintiliani, R. D., Nightingale, C. H. Platt, D, Crowe, H. Grotz, R., Orlando, R., Strycharz, C., Tetreault, J., and T. Lerer. 1991. Effect of Including Recommendations of Antimicrobial Therapy in Microbiology Laboratory Reports. Diagn. Microbiol. Infect. Dis. 14: 157-166.

7. Baron, E. J., L. R. Peterson, and S. M. Finegold. 1994. Bailey & Scott's Diagnostic Microbiology, 9th Ed. Mosby, St. Louis.

8. Edberg, SC., Trepeta, R. W. 1983. Rapid and economical identification and antimicrobial susceptibility test methodology for urinary tract pathogens. J. Clin. Microbiol.. 18: 1287-1291.

9. Ellner, P. D. 1985. A dozen ways to achieve more cost-effective microbiology. Med. Lab. Observer. 17:40-44.

10. Everett. G. D. 1983. Effect of cost education, cost audits, and faculty chart review in the use of laboratory services. Arch. Inter. Med. 143: 942-944.

11. Fan K. et al., 1993. application of rejection criteria for stool cultures for bacterial enteric pathogens. J. Clin. Microbiol. 31: 2233-2235.

12. Garcia, L. S. and D.A. Buchner. 1985. Microbiology's economics: dissecting procedures and costs. Med. Lab. Observer. 17: 51.

13. Griner, P. F., and R. J. Glaser. 1982. Misuse of laboratory test and diagnostic procedures. N. Engl., J. Med. 307: 1336-1339.

14. Gross, P. A., C. L. VanAntwerpen, W. A. Hess, and K. A. Reilly. 1988. Use and abuse of blood cultures: program to limit use. Am. J. Infect. Control 16: 114

15. Harris, P. C. Sealey, L. B. 1986. Practical cost savings in microbiology. Med. Lab. Observer 18:32.

16. Isenberg, H. D. 1992. Clinical Microbiology Procedures Handbook, Vol. 1. American Society for Microbiology, Washington, DC.

17. Kachoris, M., K. L. Ruoff, K. Welch, W. Kallas, and M. J. Ferraro. 1988. Routine culture of stool specimens for *Yersinia enterocolitica* is not a cost-effective procedure. J. Clin. Micro. 26: 582-583.

18. Kelley, M. T. 1986. Making organism identification cost-effective. Med. Lab Observer 18:55.

19. Martin, A. R., Woof, M. A., Thibodeau, L. A. Dzau, V., and E. Braunwald. A Trial of Two Strategies to

Modify the Test Ordering Behavior of Medical Residents. N. Engl. J. Med. 303 1980: 1330-1336.

20. Martin, Bettina. 1991. The CLMA Guide to Managing a Clinical Laboratory. Clinical Laboratory Management Association, Inc. Malvern Pennsylvania

21. Morris, A., J. 1992. Application of rejection criteria for stool ovum and parasite examinations. J. Clin. Microbiol. 30: 3213-3216.

22. Moss, F. M., M. W. McNicol, D.A. McSwiggan, and D. L. Miller. 1981. A survey of antibiotic prescribing in a district general hospital. I. Pattern of use. Lancet 2: 349-352.

23. National Committee for Clinical Laboratory Standards. Cost Accounting in the Clinical Laboratory; Proposed Guideline. NCCLS publication GP11-P, Pennsylvania, 1990.

24. Parson, P.E. Lichtenstein, R., and Berki, S. E. Costs of Illness, United States, 1980. In *National Medical Care Utilization and Expenditure Survey*, Series C. National Center Health Statistics Analytical Report No. 3 Pb-86-20403. Washington, D. C.:U. S. Government Printing Office, 1986.

25. Peters, C.S., Hernandez, L., Sheffield. 1988. Cost containment of formalin-preserved stool specimens for ova and parasites from outpatients. J. Clin. Microbiol. 26:1584

26. Robinson, A. 1994. Rationale for Cost-Effective Laboratory Medicine. Clin. Microbiol Rev. 7: 185-199.

27. Schentag. J. J., C. H. Ballow, A. L. Fritz, J. A. Paladino, J. D. Williams, T. J. Cumbo, R. V. Ali, V. A. Galletta, M. B. Gutfeld, and M. H. Adelman. 1993. Changes in Antimicrobial Agent usage

Resulting from Interactions Among Clinical Pharmacy, the Infectious Disease Division, and the Microbiology Laboratory. Diagn. Microbiol. Infect. Dis. 16:255-264.

28. Sharp, S.E. 1993. Effective reporting of susceptibility test results. Diagn. Microbiol. Infect. Dis. 16: 251-254.

29. Siegel, D. L. 1990. Inappropriate testing for diarrheal diseases in the hospital. JAMA 263:979-982.

30. Smith, J.W. MD, 1985. The Role of Clinical Microbiology in Cost-Effective Health Care. 1984 CAP Conference. Published by the college of American Pathologists, Skokie, Illinois 1985.

31. Studnick, J., D. D. Brandham, J. Marshburn, P.R. Foulis, and J. V. Straumfjord. Measuring the Impact of Standing Orders on Laboratory Utilization. Lab. Medicine. 23: 24-28, 1992.

32. Summanen, P., E. J. Baron, D. M. Citron, C. A. Strong, H. M. Wexler, and S. M. Finegold. 1993. Wadsworth Anaerobic Bacteriology Manual. 5th Ed. Star Publishing Company, Belmont, CA.

33. Trepeta, R. W., Edberg, S. C. 1984. Methylumbelliferyl-beta-D-Glucuronide-based medium for rapid isolation and identification of *Escherichia coli*. J. Clin. Microbiol.: 19: 172-174.

34. Williams, D. N., Lund, M. E., Blazevic, D.J. Significance of urinary isolates of coagulase-negative *Micrococcaceae*. J. Clin. Microbiol. 1976: 3: 556-559.

35. Windelman, J. W. 1985. Quantitative analysis of cost-savings strategies in the clinical laboratory. Clin. Lab. Med. 5: 635-651.

36. Young, D. S., and G. J. Tidwell. 1989. A hospital-wide task force to reduce lab utilization. Medical Lab. Observer. 21:31-40.

CHAPTER IV

COST CONTAINMENT IN THE CLINICAL IMMUNOLOGY LABORATORY

Rhonda Emerick, B.S., MT (SM)

Ronald J. Harbeck, Ph.D.

Diagnostic Immunology Laboratory
National Jewish Center for Immunology
and Respiratory Medicine
Denver, CO 80206

I. INTRODUCTION

Enormous progress has been made in the field of immunology over the last three decades. Not only have important advances been made in our understanding of the basic mechanisms of the host response to foreign elements, but also of the clinical application of these mechanisms.

Historically, the field of medical laboratory immunology has evolved from the fields of infectious diseases, serology, and immunohematology especially as it relates to traditional blood banking practices.

Now diagnostic immunology permeates many subspecialties of medicine, and today's diagnostic immunology laboratory serves a heterogeneous mix of medical specialties and subspecialties.[1] These subspecialties rely on immunology laboratories to help make a diagnosis or monitor treatment of patients, and include, but are not limited to, allergy, dermatology, infectious diseases, rheumatology, pediatrics, gastroenterology, hematology, pulmonology, nephrology, and endocrinology.

Immunology testing in the past was often carried out in other sections of the clinical laboratory, e.g., the chemistry laboratory, microbiology laboratories, and/or the blood bank. There is now a trend among large medical centers to centralize all immunology testing in one section.[2]

Today the typical components of an immunology laboratory are cellular immunology (lymphocytes and phagocytic cells) and humoral immunology (immunoglobulins and antibodies). Histocompatibility/tissue typing and tumor markers are relatively recent additions to some immunology laboratories. However, the range of services provided by any single laboratory depends on the nature of the institution where the lab is located. Its been said that no two immunology labs are the same.

There are now a plethora of tests carried out in an immunology laboratory which evaluate the various components of the immune system. The immunology laboratory will continue to grow as basic scientists unravel the interactions of the various components of the immune system, and delineate the role of these components in the immunopathogenesis of disease. One can expect that a marked expansion of the immunology laboratory's test menu will be increasing in the future with the addition of assays for cytokines, adhesion molecules, further applications of flow cytometry, and molecular diagnosis.[3]

Are there ways that the immunology laboratories can control costs to meet the changing fiscal challenges? Today's immunology laboratory must become a highly efficient service provider in order to support its client's needs, and to survive in the competitive marketplace. Wise decisions, development of management systems that can choose the most cost-effective methods and schedules, and the identification and elimination of non-essential services are important ways of meeting these challenges.[4,5]

Laboratorians must identify opportunities in the marketplace, explore new pricing strategies, be flexible, and initiate aggressive marketing efforts. Often these are mechanisms that laboratories have not explored in the past.

Mergers of hospital facilities and consolidation of laboratory services are important components of cost accounting that laboratory administrators are exploring in order to control costs. There is a need for administrators and clinicians to work with the laboratory in controlling the cost of the laboratory operations. It behooves laboratorians to be aware of forces that can give rise to changes in the way we laboratories do business and to plan accordingly.

Prospective payment for immunology laboratory services, as with other laboratory services, has turned the laboratory from a revenue center into a cost center. Because of this, and other pressures from managed care entities, the immunology laboratory will have to be more aware of reducing costs since revenue increases are unlikely. Laboratories must become low cost, high service providers in order to support managed care initiatives and to survive in the competitive marketplace. Fixed and variable expense projections, indirect cost allocations, utilization of laboratory services, capital requirement estimates, and direct and indirect test cost profiles are all areas that need to be explored and analyzed.

Under the Clinical Laboratory Improvement Act of 1988 (CLIA '88) many of the procedures performed in the immunology laboratory are classified as moderate or high complexity. Because of this, the personnel and supervisory requirements are more specific and can be more costly. Some predict that these new regulations and the changing healthcare environment will decrease the demand for the highly complex and esoteric immunology laboratory tests.[6] Because of increase in costs, many of the moderate and complex immunology procedures that do not require a rapid turnaround time (TAT) may be shifted to a reference laboratory.

When examining the cost containment opportunities in an immunology laboratory, one must understand that the immunology laboratory has some unique aspects which distinguishes it from most other laboratory sections.

While chemistry laboratories often batch testing, and perform analysis on dozens if not hundreds of samples at one time, this is often not practical or possible in the immunology laboratory. The immunology laboratory may be called upon to perform only a single test, e.g., a cellular assay such as neutrophil chemotaxis or lymphocyte mitogenesis. Thesc tests are very labor intensive, and have the added cost of setting up an equally labor intensive control in parallel. Economy of scales would apply in these examples, but opportunities to run more than one of these tests at a time may be infrequent. Testing in an immunology laboratory generally is more hands-on, and uses a more traditional "wet lab" approach. Reducing the TAT of labor intensive, high-tech procedures especially when volumes are low is challenging.

New technologies are emerging, but often these may add to the cost rather than contain costs in the immunology lab. Another factor which can increase the operating cost of an immunology laboratory is the training of personnel. Specialized training in immunology is not often taught in medical technology schools. Technologists often require on the job training, and the time needed to teach a new technologist new assay procedures can be considerable and expensive.

II. AUTOMATION IN AN IMMUNOLOGY LABORATORY

Automation is being introduced in many areas of the laboratory which traditionally have not had it, such as microbiology, blood banking, and in some cases immunology. In most areas of the laboratory, automation has resulted in significant cost savings by requiring fewer people with less skills, providing more rapid throughput of assays, significantly reducing reagent costs, and providing more reproducible and accurate test results. In addition, it has been said that automation does not carry a benefits package, i.e., one does not need to pay for overtime, insurance benefits, holiday pay, etc.[7] The expense of a new technology or instrument can often be justified to administration if it can be shown that productivity will increase.

Unfortunately, in many immunology laboratories automation is not always an option. Automation in immunology laboratories has been slow in coming because of the array of testing, the multiple steps involved in many of the procedures, and the relatively low volume of some tests.

Some of the procedures in an immunology laboratory, such as those involved in looking at cellular functions, are nearly impossible to automate because of the sterility requirements, the need to examine the functional activity of viable cells, and the isolation steps that do not lend themselves easily to automation. However, when examining serum components in contrast to cellular components, automation and advances in technology have aided in reducing costs and labor.

Devices such as automatic dilutors and ELISA microtiter plate washers, that can provide savings in labor and expedite TAT of tests, have been developed and have applications in larger immunology laboratories with greater test volumes. These new automated devices provide more consistency of technique in the reagent addition and washing steps used in enzyme immunoassays (EIA) procedures, which are critical in these types of assays. With increased consistency, there are fewer repeat tests due to technical error, thus the TAT, accuracy, and reproducibility of test results are greatly improved. All these factors contribute to cost containment and increase productivity in the clinical immunology laboratory.

Additionally, the use of plate washers allows for less hands on time for the user as well as being faster and safer with the elimination of potentially biohazardous aerosols. Many of these units will allow the user to program for number of washes and volumes of wash to use per well. There is also the option of having a plate stacker, which allows multiple plates to be processed, making this system truly "hands off".

In many cases robotics has enabled the immunology laboratory to be more productive. For example, in the performance of serological assays, automatic dilutors free the technologist from performing endless numbers of serial dilutions of serum which traditionally have been the hallmark of many serological procedures.

These automatic dilutors can be a great asset to the immunology lab in many ways. Dilutors provide freedom to the technologist to perform other tasks while the samples are being processed. Also, the degree of accuracy and reliability is greatly enhanced by using an automated system. Many of these systems provide the user with a wide range of versatility and flexibility. The computer software allows the user to program the instrument to perform the pre- and subsequent serial dilutions, obtain sample worklists, add timed reagents, and operate in the batch mode as well as provide barcode reading capability.

This type of automation also adds to the productivity of the immunology laboratory by eliminating physical difficulties which plague many immunology technologists due to the constant repetitive motion needed in pipetting.

The automatic dilutor also allows for a safer working environment because the technologist has less contact with the specimen and reagents since fewer aerosols are generated. In many of the EIA procedures the substrate is a potential or known carcinogen which is of danger to the technologist. The automatic dilutor reduces this danger.

However, a disadvantage of this type of automation is that it requires space for both the instrument as well as a work space for the technologist. This is especially true of the dilutor which must have a computer to operate it. The cost of this type of instrumentation may be prohibitive, but it is often possible to obtain these types of instruments through a lease agreement with the manufacturer.

There are a limited number of manufacturers who have developed "fully" automated systems. These systems are truly walkaway. The only contact the user has with the system throughout the run is to load the reagents, samples and controls, and program the system. Once all of this is done, the system needs no further attention. Upon completion of the assay run, the results are calculated. The real advantages to using this type of automated system are obvious: increased productivity and throughput, increased precision and accuracy, increased worker safety, and less operator error from sample misidentification, transcription, and calculation.

The trend for manufacturers will be to continue to develop instruments that will be smaller, more sensitive, less costly, and that require smaller volumes of reagents and samples and less technical input. These new instruments will result in greater reproducibility and stability of the assay systems and improved TAT. CLIA '88 will influence manufacturers to design instruments that will remove tests from the moderate and high complex levels to the less complex levels.

III. INSTRUMENTATION

While in the past, automated equipment could be expected to have a life-span of five years, today, better electronics and circuitry are increasing their life-expectancy. However, technologies change so rapidly that within 5 years a system can be appreciably outdated. New instrumentation can decrease costs of service contracts, down time costs, maintenance, and troubleshooting costs. New instrumentation may also decrease reagent costs, technologist time, service expenses, and increased costs associated with sending the sample out.

Competition among manufacturers is at times fierce, so it is wise to contact as many manufacturers' sales people as possible, and let them know you are "shopping around". This healthy competition between manufacturers can serve the laboratories interest and can ultimately result in costs savings.

Often, instrument manufacturers are aware of demo or refurbished instruments.[8] These are sold at a fraction of the cost of new instruments. Often, reconditioned instruments come with full service contracts and warranties. Training opportunities frequently come with the purchase of a reconditioned item, either from the vendor selling the item or from the original manufacturer. Some vendors obtain instruments as a result of a trade-up situation or through a leasing company. Since these vendors pay only a fraction of the new cost price, they are able to pass these savings onto the laboratory. A careful check of references and conditions of the sale are important in any transaction of this nature.

Some manufacturers will provide reagent rentals and leases of instruments. Many different types of financial arrangements are possible, and one should allow the manufacturer's sales representative to lead you through each of their different options.

Often, it makes more sense for a laboratory to invest in new instrumentation, particularly if it can be shown that workload consolidation will result in more efficient handling of specimens and improved TAT. Random access instruments are currently available that allow consolidation of immunoassay testing. In a recent article, the supervisor of one large hospital describes how he was able to generate new revenue sources, and remain competitive in the health care market by acquiring a random access system and consolidating workloads.[9] Whereas previously, several analyzers were required to perform TDMs, CK-MBs, thyroid profiles, cortisols, ferritins, CEAs, PSAs, hCGs, rubellas, etc., the acquisition of a single analyzer with the capability of performing all these tests improved productivity, enhanced revenue, and resulted in a cost savings to the laboratory. The features that this author considered when purchasing a new instrument included random access, walk-away automation, low maintenance, ready-to-use, refrigeration, on-board reagents, low cost per test, bidirectional interface capability, broad test menu, good calibration stability, and strong customer support.

Laboratory personnel need to keep abreast of new developments in testing or instrumentation in immunology by reading the literature and attending meetings where manufacturers and reagent providers show their products. This type of "one-stop shopping" can be a real time saver for the laboratory.

IV. ASSAY KIT CONFIGURATION

In addition to advances in the technology to automate physical tasks, i.e., pipetting and washing, many companies have developed products that are more user friendly. These have included enzyme immunoassay kits that enable the laboratory to customize the panels. This in turn allows a laboratory to perform tests in batches. Batch testing was virtually impossible for many of the tests performed until recently. Now, a greater selection of antibody and antigen detection kits are available, with many more types being introduced every year. Autoantigens, parasitic and fungal antigens, or antibody detection systems are just a few of the kits currently available.

Reagents such as the wash buffers, substrates, conjugates, and stop solutions are interchangeable when used with other test kits manufactured by the same company. The procedures for many of the various tests offered by the same manufacturer are also the same.

In many cases, a laboratory can purchase separate interchangeable reagents at costs which may be considerably less than the purchase price of an entire kit. For example, the same conjugated antibody can be used with a variety of antibody detection systems, while one employs separate, test-specific antigens. These factors aid the technologist by providing ease of use, continuity, fewer separate protocols with separate reagents to remember, and ease in teaching procedures to new employees or users.

"Break away" wells, coated with antigen or antibody by the manufacturer, allow the user to batch several different tests within the same assay run. This is an advantage to the immunology lab in several ways: it is more cost effective, provides improved TAT (since the technologist can set up a single run to measure various antigen-antibody systems instead of multiple testing), and allows the lab to obtain better volume discounts from the manufacturer. Similar potentially low volume tests can be batched together to make one large run.

The savings are noticed in labor when one run is being performed, instead of two or more small runs performed on separate occasions. When one run is performed, rather than two or more small runs, there is a further savings because less substrate, wash and stop solutions are used. Also, less time is spent on interpretation of data and computer entry when one large run is performed. Referral accounts and physicians are satisfied because low volume tests are easy to add to other tests of moderate volume, thus improving TAT.

Another area where technological advancement has benefited the immunology lab is in measuring of the levels of various serum components, e.g., immunoglobulins and complement proteins. As with most lab tests, it may be important to the clinician, for diagnostic purposes, to obtain the results of IgG, IgA, and IgM, or the complement components, C3 and C4, levels quickly and accurately. In the past many of these protein levels were measured by radial immunodiffusion. In this technique a patient's serum is added to a well of an agar containing an antibody directed against the component one wishes to measure. The size of the precipitin ring that forms around the well is proportional to the concentration of the antigen present in the serum. This technique usually requires an incubation of two days. Ring diameter is measured which can be somewhat subjective. The sensitivity is generally not very good relative to newer assays nor is the quantitation very accurate.

The advent of nephelometry, an instrument which measures light scatter as antigen-antibody complexes are forming, makes quantitation more precise and sensitive and improves TAT. The earlier nephelometry systems, however, still required a great deal of technologist involvement. Only one sample with one protein could be tested at a time. Today's nephelometers can perform multiple analytes on multiple samples at any given time. These systems provide walk away capability which has significantly reduced labor costs and increased productivity. Nevertheless, RID is a technique which may still be applicable to small labs because the volume of samples do not justify the expense of instrument acquisition.

Many of the automated systems can be interfaced with the laboratory information systems to eliminate tedious entry of data into the computer. Interfacing is especially helpful in reducing clerical errors from manual entry. Monitoring quality control can be a problem with a large number of manual tests because tests are hard to track, and additional time is needed to enter results into a computer for evaluation.

Automation in the immunology laboratory has made a significant impact in this particular area because much of the computer/software has been developed to evaluate quality control (QC) directly as it is run. This is a time savings for the user because there is instantaneous feedback as to the acceptability of controls and calibrations. This information is also easier to retrieve for instrument and quality control troubleshooting review.

The computer software programs used in automation also lend themselves to better record keeping when it comes to maintenance, quality control, and linearity data. This detailed record keeping is required for accrediting agencies such as CLIA, CAP and JCAHO.

The easy access to this information with computer programs allows not only for better quality assurance measures, but saves personnel endless hours that were spent in the past trying to retrieve this information needed for completing reports and proving that the QC, etc., were within acceptable established ranges.

The utilization of the EIA test has not only provided some form of automation to the immunology lab, but has improved the sensitivity and specificity as compared to previously used procedures such as immunofluorescence. This use of EIA has also allowed the immunology laboratory to report out more quantitative answers, giving the clinician more accurate results on which to base decisions. Thus, automation has contributed to more precise results in a more timely manner.

V. TECHNOLOGICAL CHANGES

The measurement of lymphocyte surface markers such as CD4 is an example of a change in technology which made cellular measurements more cost effective with less technologist involvement. The processing of lymphocytes for surface immunophenotyping has gone from a complicated, labor-intensive, and time-consuming process to a relatively rapid, automated procedure, i.e., the whole blood staining technique and lysis.[10,11]

Up until a few years ago, the steps involved in separating peripheral blood mononuclear cells on a density gradient material included: washing the cells, incubating the isolated cells with an appropriately unlabeled monoclonal antibody to the surface marker one was interested in, washing the cells again, and labeling the cells with an appropriately tagged secondary antibody. Next, using fluorescent microscopy, a technologist would enumerate the number of fluorescent cells in a field of mononuclear cells. Generally, less than 300 cells were counted. As with most fluorescent based assays, subjectivity in the reading was great. Furthermore, the time involved in testing a few blood samples for the appropriate phenotypes can take a significant part of the work day.

Today, the whole blood lysis technique is preferred, not only because it avoids the potential differential loss of specific cellular populations resulting from centrifugation techniques, but also it is less labor intensive, more rapid, and requires less blood. With this technique, directly conjugated cocktails of monoclonal antibodies are added to small (100 µl) volumes of whole blood, and incubated for a short period of time (generally less than 30 minutes). The red cells are lyzed, and the cells fixed and preserved. The percentage of cells expressing the various markers, e.g., CD3, CD4, and CD8, in the samples can then be determined by a flow cytometer.

Manufacturers have developed automated flow cytometry systems which will carry out the steps much more rapidly, with thousands of cells analyzed within minutes.

Flow cytometry is a rapidly developing technology which has moved from the research laboratory into the clinical laboratory. The multitude of steps involved in processing the blood have been shortened considerably resulting in fewer errors and considerable savings of technologist time. Today dozens of samples can be set-up and analyzed in a single work day.

Obviously the HIV epidemic and the need for physicians to obtain CD4 levels in a timely fashion and in an accurate and reliable manner stimulated manufacturers to develop the necessary instrumentation, methods, and reagents so that this can be accomplished rapidly and efficiently.

Often, in a community hospital and laboratory, the low test volume does not justify the investment of the above instrumentation. Analytical flow cytometers, for example, are priced in the $75,000 to $125,000 range. Also, a high level of technical expertise to operate a flow cytometer and to interpret the results is required. Considering that many reference laboratories offer this testing at prices that laboratories with low volumes can't meet, it may make more sense to send these tests out.

Recently, some manufacturers have released relatively low priced instruments which will automatically perform immunophenotyping for the common lymphocyte subsets, e.g., CD3, CD4 and CD8, giving absolute counts of these subsets from a single sample. These instruments require little or no operator intervention. They are designed for smaller laboratories in which the volume of immunophenotyping does not justify the expenditure and training associated with the more expensive flow cytometers, or for laboratories involved in measuring only CD3, CD4, and CD8 cells. Other technologies, including EIA based methods, are being introduced for enumeration of lymphocyte subsets as an alternative to the high cost of reagent and instrument purchase.

VI. SELECTION OF APPROPRIATE TESTS AS A COST SAVING METHOD

Selection, by the physician, of an appropriate laboratory test for a particular diagnosis is of critical importance in containing health care costs. It is frequently the responsibility of the laboratory to educate the medical staff as to the use of screening procedures as well as subsequent cost-effective procedures which will aid in making a diagnosis.

Screening tests should always be set up before more specialized tests are performed. This is of particular importance in an immunology lab where test costs can be steep!! For example, when examining a patient's neutrophils for their ability to undergo oxidative metabolism, a relatively simple, inexpensive nitroblue tetrazolium (NBT) dye reduction test will detect carriers and patients with chronic granulomatous disease (CGD). This provides as much information to the physician as the more labor intensive and costly tests such as superoxide production, *in vitro* bactericidal assays, and neutrophil chemiluminescence.

For the diagnosis of primary immunodeficiency diseases, the physician can obtain some relatively simple tests instead of the more complex, esoteric, and costly procedures.[12,13] One can consider the host defense system as four interactive compartments, i.e., complement, phagocytic, lymphocytic and antibody. Screening tests should allow one to determine which of the four arms is defective. A description of some easy, cost effective tests for ruling out immunodeficiencies, is shown in Table 1.

It is important when selecting assays to evaluate immunodeficiencies that they be broadly informative, relatively simple to perform, and cost-effective. Once the defective arm of the host defense system has been detected, more specialized immunology laboratories (often in research settings) can be used to pinpoint the defect. It is imperative that the proper choice of laboratory tests for screening purposes be made since many of the secondary tests can be very costly.

TABLE 1.

Disorder	Screening tests
All immunodeficiencies	CBC, examination of red blood cells for Howell-Jolly bodies, ESR.
Antibody deficiency	Protein electrophoresis, immunoglobulin quantitation, and specific antibody responses.
Complement deficiency	CH50 assay
Cell-mediated (T cell mediated) deficiency	Absolute lymphocyte count, delayed hypersensitivity skin tests.
Phagocytic cell deficiency	Absolute neutrophil count, nitroblue tetrazolium test (NBT).

When screening for any suspected immunodeficiency, a complete blood count and an erythrocyte sedimentation rate (ESR) can provide valuable information. A normal ESR will rule out a chronic, bacterial infection while a normal platelet count will rule out Wiskott-Aldrich syndrome. Severe T-cell defects can be ruled out if the patient has a normal lymphocyte count although it must be remembered that infant's lower limit of the normal reference range for lymphocytes is around 4500/mL. If the neutrophil count is normal, congenital and acquired neutropenia and some other neutrophil functional disorders, such as leukocyte adhesion deficiency, are eliminated. The absence of Howell-Jolly bodies in red blood cells helps to rule out congenital asplenia.

For evaluating antibody deficiencies, screening tests such as serum protein electrophoresis can rule out agammaglobulinemia and hypogammaglobulinemia. This can be further evaluated by examining the serum for the major immunoglobulin isotypes and for specific antibody activity.

Most complement deficiencies can be ruled out if the CH50 test is normal. In the CH50 assay, complement activity is measured by the addition of a patient's serum to a suspension of antibody sensitized sheep red blood cells (SRBC). The CH50 is an arbitrary unit defined as the amount of complement needed to lyse 50% of the SRBC. The results are expressed as the reciprocal of the serum dilution giving 50% hemolysis.

Some simple screening procedures to assess the integrity of the T cell mediated arm of the immune system, often called cell-mediated immunity, are the absolute lymphocyte count and a delayed type hypersensitivity skin test.

For evaluating the phagocytic cell arm, an absolute neutrophil count along with a test to measure the stimulation of the neutrophil respiratory burst, the nitroblue tetrazolium dye reduction test (NBT), can be performed.

In performing any cellular assays, four points must be remembered 1) they are time consuming, laborious and expensive assays, 2) they are biological assays and hence are highly variable which may require multiple replicates and possible repeat testing, 3) they often need to be set up as soon as possible after obtaining the blood due to lability of the cells and 4) the interpretation of the results must be made conservatively.

Indirect immunofluorescence is one of the more commonly performed tests in an immunology laboratory. It is frequently used in the detection of autoantibodies in various autoimmune diseases. This technique has been used to detect autoantibodies against a variety of cellular targets including nuclear, mitochondrial, and cytoplasmic antigens such as anti-neutrophil cytoplasmic antibodies (ANCA) and thyroid epithelial cells. The most popular of these is the antinuclear antibody (ANA) assay which can be used to screen for systemic lupus erythematosus and other connective tissue diseases. Many kits are commercially available using cultured human epithelium (Hep-2) and are relatively inexpensive. Hence this additional test can be utilized in an immunology lab to screen for nuclear antigens, i.e., SS-A, SS-B, Sm, and RNP.

VII. SENDING TESTS TO REFERENCE LABORATORIES

To send out to a reference laboratory or not? When is it justified to perform your own testing. The use of a reference laboratory is an option that most immunology laboratories need to explore. When deciding on a reference laboratory, you must ask yourself the following questions: 1) is the requested test of low volume requiring considerable expense to set up infrequently, (higher volumes are the key to the game!!! It is not cost effective for any laboratory to perform tests in low-volumes), 2) will I improve my TAT for the client, 3) is the integrity of the sample compromised by sending it to a remote reference laboratory, 4) are the training costs excessive 5) is the loss of control of a test important to my laboratory and institution?

In our laboratory the main driving force for deciding whether to send a particular test to a reference lab is volume. The large majority of immunology tests that are sent out are done so because there is a very low number being ordered, and the cost of reagents and labor do not justify performing the assay. Some examples of tests that have been sent out in our laboratory are anti-MPO and PR3 antibodies and staph IgE binding assays.

Other tests with moderate volume may be going to a reference lab because instrumentation, reagents, and technical expertise are not available in our facility. Some examples might include PCR and Western blot procedures.

Many reference laboratories will provide courier services to transport samples back to their home base for testing at no additional cost to the client. This can be a cost savings for the immunology lab. In addition, reference labs that receive a large percentage of the "send out tests" are willing to put a computer system in place that allows the sending lab to enter samples and retrieve results directly from their system. This is an advantage to both parties by eliminating requisition entry by the reference lab and faster TAT of results to the client. On line computers with instruments add to the cost savings and decrease errors by lessening the personnel interactions in the reporting of results. Thus, with this computer based technology, reference labs can sometimes provide faster TAT, at a lower cost, for tests that are infrequently ordered.

However, some tests are so esoteric and infrequent that the choices of labs may be limited. In this case, cost and TAT may be irrelevant because choices of reference labs are limited. However, our experience has been that most reference laboratories will gladly provide information on where a test can be done if they themselves don't do the assay. Another excellent source for finding laboratories that may perform esoteric tests is the American Association for Clinical Chemistry's (AACC) publication *DORA '94-96: Directory of Rare Analyses* (AACC Press, 2101 L Street, NW, Suite 202, Washington, DC 20037).

When choosing a reference lab, one should keep in mind all of the following: cost per test, methodology, TAT, and level of service. It's important that the reference lab's technicians and directors be accessible to solve problems and answer technical concerns in an accurate and timely manner.

VIII. THE DESIGN OF THE CLINICAL IMMUNOLOGY LABORATORY

Some things that are not often considered in saving costs in a laboratory is the appropriate architectural design as related to the function of the laboratory[2]. In laboratories where there are space problems, one should look at the placement of equipment to be sure that it is appropriate to the activity. Often, equipment is placed wherever one can find room, without regard to its use in the flow of production in the laboratory. Careful thought and planning of the laboratory space is critical to optimizing a lab's productivity and efficiency.

In our laboratory, we found that the technologist was using a laminar flow hood, a centrifuge, and a scintillation counter, all at distant sites from one another for lymphocyte stimulation tests. Because each of the pieces of equipment was originally purchased for different or multiple purposes, they had remained at these sites because of space problems.

Many times, when placing equipment and expanding work space, we forget to consider the prospect of more growth and an increase in workflow, and what impact these will have on the work environment. If careful thought to adequate space is not allowed when designing a lab, then the flow of the whole department may be upset by the addition of one new piece of equipment which may be purchased if increasing test volume leads to further automation.

It is also important to have adequate phones and computer terminals as well as adequate desk top space for writing in notebooks. How the workspace is arranged can also contribute to safety of employees. The safety office at one's institution can help to provide ideas or suggestions on how best to utilize existing space for optimal safety.

Laboratorians often become indifferent to the layout of their laboratories. Asking a colleague from another lab to spend a day in one's laboratory to observe the work flow, could provide useful ideas for restructuring the laboratory. If the opportunity to develop a new work space is available, keep in mind the potential for growth and for the type of growth that may be foreseen. This will help you develop a work space that will be as functional in 10 years as it is today.

Another alternative to maximize laboratory work flow is to employ an architectural firm with experience in the design of clinical laboratory facilities. As the authors can attest from personal experience, these firms can provide many valuable ideas to assist one's operation, even if major renovations or the construction of a new laboratory is not an option.

IX. SPECIFIC TEST MODIFICATIONS TO CONTROL EXPENSES

The personnel of an immunology laboratory should continually strive to modify current procedures to reduce cost and labor, and at the same time provide a safer working environment with minimal exposure to hazardous materials.

Adequate validation of changes in test procedures are needed before adapting any new procedure. The following are some examples of commonly performed tests in the authors' laboratory that have been modified to be more cost effective, safer, and faster.

One assay where the greatest number of changes have been implemented is the lymphocyte proliferation test to mitogens, antigens, and beryllium salts.[10,14] This assay is a very common procedure, used both in immunology research as well as in clinical immunology laboratories, to assist in the diagnosis of immunologic disorders. Lymphocyte proliferation to beryllium salts is used to assess chronic beryllium lung disease, often in an occupational setting. This type of assay is very labor intensive and takes several hours to perform.

At one time, many of the procedures requiring the use of whole blood to isolate lymphocytes or neutrophils had to be collected into a heparinized syringe, and testing had to be started within hours of collection. As a result of our research, we determined that comparable results were obtained from blood collected in heparin tubes which was allowed to remain at room temperature and tested within 24 hours.

A similar change was made in the NBT assay. In this procedure, glass adhering neutrophils are activated with agents which induce oxidative metabolism within a normal cell. When the NBT dye is reduced, it yields a blue formazan precipitate which is easily visualized within the cytoplasm of the cell undergoing oxidative metabolism.

In the past, blood was collected from the patient and placed immediately on previously prepared coverslips and stained. After a period of experimentation, we found that the sample could be collected in heparin, and the test performed within 24 hours with no compromise of test results. This simple modification has allowed the immunology lab to conserve labor by batching two or more patients whenever possible. The time savings in labor can add up to several hours per patient depending on the assay being performed.

Discovering that blood samples for neutrophils and lymphocyte studies could be held for up to 24 hours has allowed our lab to perform testing on patients from other states and on patients who are unable to travel to our facility and are drawn elsewhere. This modification has allowed us to provide better service to our clients while keeping costs down.

Other changes made include the use of commercially prepared density gradient tubes (Accuspin Systems - Histopaque - 1077, Sigma Diagnostics, St. Louis, MO) for the isolation of mononuclear cells for subsequent use in lymphocyte proliferation assays. These tubes are conical centrifuge tubes containing a density gradient solution with a plastic filter at the top of the solution.

Heparinized blood can be poured directly into the tube eliminating the careful layering necessary when using density gradient solutions. In addition, after centrifugation, the plasma layer can be removed and the remaining white cell layer can be poured directly into another tube for washing of the cells.

Mononuclear cell recovery is improved with this system, and the tedious task of carefully removing the mononuclear cell layer using the old method has been eliminated. In addition to the standard 50 ml size, these tubes are also available in a 15 ml size for use with pediatric patients. These tubes are more expensive then preparing density gradient tubes in-house, but the savings in labor costs on large volumes of cell separations justifies the expenditure.

Our laboratory described a method utilizing Percoll (Pharmacia) for the recovery of neutrophils and lymphocytes from small volumes of blood, e.g., pediatric populations.[15] With this method, the use of a single discontinuous density gradient tube allows for the separation of both granulocytes and mononuclear cells. The separation procedure does not influence the functional activity of the neutrophils or the mononuclear cells. Hence, in cases where one needs to assess both neutrophil and lymphocyte functions from a small amount of blood, this technique should be considered.

Chemotaxis assays of polymorphonuclear neutrophils (PMN) are usually carried out with cells that have been isolated from density gradients. This is generally a time consuming procedure.

Two recent publications have described modification of chemotaxis assays which leads to an improvement in the TAT of this assay, and appreciably reduces the cost. In one paper, chemotaxis was carried out using unseparated cells.[16] Whole blood at 100%, 50%, and 25% was compared to separated cells in chemotaxis using *N*-formyl peptide and interleukin-B as chemoattractants and a Boyden-type chamber with polycarbonate membranes. It was found that PMN's from whole blood exhibited a greater chemotactic activity.

In another paper the authors described a chemotaxis assay using a system consisting of a 96-well chamber, polycarbonate filter membranes, and a 96-well microtiter plate.[17] PMN peroxidase activity was used as a measure of the PMN migration. Detection of peroxidase was determined using a standard ELISA plate reader. Up to 80 samples along with 16 controls can be set up in 2 hours. The authors suggest that the rapidity of the test, the convenience of handling, and the use of standard laboratory equipment result in a low cost assay.

In the lymphocyte stimulation assays, cells are cultured for several days, ^{3}H-thymidine is added as an indicator of new DNA synthesis, and this is followed by harvesting the cells onto filter paper mats for counting of the incorporated ^3H-thymidine. Cell harvesters are now available which will harvest all 96 wells of a microtiter plate at one time. The harvesters cut labor costs appreciably while allowing for fewer manipulations of the cells. In addition, radioactive counters are commercially available which are able to count the samples from a 96 well plate simultaneously. These counters do not use scintillation fluid which eliminates the problems associated with disposal and handling of biohazardous fluids.

Another avenue worth exploring in labs performing cellular assays requiring radioisotopes is the use of non-radioactive alternatives. Several non-radioactive techniques are now available for use in proliferation assays, cytotoxicity studies, and other assays in which isotopes have traditionally been used.

Cellular proliferation[18] has been measured by cleavage of the yellow tetrazolium salt, 3-(4,5-dimethylthiazol-2-yl)-2,5-diphenyltetrasolium bromide (MTT) to a colored formazan product by mitochondrial enzymes in viable cells. In addition, this salt has been used in studies to examine cell stimulation, cytotoxicity studies, and cytokine production. There is a recent study reporting on an improved MTT assay giving greater reproducibility and accuracy.[19]

A new one-step non-radioactive assay has been described to determine the proliferation of murine lymphocytes.[20] This procedure uses Alamar Blue dye added to cultures. The degree of color change is reflected in the extent of cellular proliferation, and can be easily measured with an ELISA reader. Results have been shown to be equivalent to ^3H-thymidine incorporation assays. One of the advantages of this assay is that proliferation can be evaluated on a daily basis, since there are no lysing, extracting, or washing steps involved. Hence this procedure offers many advantages, including non-radioactivity, simplicity, less cost, non-labor intensive methodology, rapid assessment of proliferation in large numbers of samples, and non-toxicity.

In addition to the safety issues associated with the use of radioisotopes, the cost of keeping and disposing of radioisotopes can be significant. It remains to be seen if this assay can be adapted to the use of human cellular proliferation studies.

Culture media for use in *in vitro* cellular assays is less expensive if bought in the powdered form and then reconstituted in one's own lab. However, one must weigh the personnel efforts and costs of preparing the media, i.e. rehydrating, sterilizing, aliquotting, etc., to the cost of ordering it pre-prepared. In addition, it may be cheaper in the long run to rely on the manufacturer's quality control of the product.

We have also found that mitogens and antigens for use in lymphocyte proliferation assays can be prepared with at least a 7 day expiration date. This has been a cost savings in labor and reagents since less time and money are spent making fresh reagents each time the assay is set up.

Another change to the lymphocyte transformation assay is the elimination of the cell separation step altogether. Lymphocyte proliferation studies using anticoagulated whole blood have been described.[21] The advantages of this technique include the requirement for less patient sample (as little as 1 ml vs. 20 mls), shorter set up time because no cell separation or washing is necessary, and fewer reagents and supplies. This whole blood method has been useful for pediatric patients and patients from whom only a small amount of blood can be obtained. The whole blood lymphocyte assay can be set up in as little as an hour, compared to up to four hours using the older procedure.

New applications for the use of flow cytometry have recently been released by Becton Dickinson. The company claims that the FastImmune™ Assay System can directly assess individual T-lymphocyte subset responses to various stimuli within a short period of time using three-color immunofluorescence.[22] In essence, the kit contains CD69 which is an early activation antigen found on lymphoid cells. In this procedure, whole blood is incubated for 4 hours with antigenic or mitogenic stimuli, the cells are stained, the RBCs lysed, the lymphocytes fixed, and the cells analyzed by flow cytometry.

The assay utilizes fluorescence triggering on CD3-positive lymphocytes and subsequent two-color analysis of the activation marker CD69 along with the subpopulation markers such as CD4 ad CD8. T-cell activation is measured as a function of the percentage of T-cell subsets that express CD69. This is a very rapid, easy procedure which gives the results in less than 5 hours. It is claimed to give results comparable to the more traditional *in vitro* lymphocyte stimulation assays which can take from 3 to 7 days (depending on the stimulus).

Furthermore, this procedure offers many other advantages over the current *in vitro* functional assays in use. There is no need to separate the mononuclear cell fraction; no tissue culturing is necessary which eliminates the need for incubators, isotopes, and other reagents and supplies; biohazard risks are reduced; multiparameter analysis is achieved; and the procedure lends itself to easier automation. This procedure could have a broad range of applications including monitoring autoimmune diseases, transplantation rejection, chemotherapy, and immune status in HIV infections. It should be noted that this test kit is for research use only and is not as yet for use in diagnostic or therapeutic procedures.

Cytokine measurements are becoming important components to measure during the immune response. Cytokines are produced by a variety of activated cells. In the past their measurement has been made using bioassays, molecular, and ELISA techniques. Recently, Jung et al.[23] using three-color flow cytometry described a technique to measure intracellular cytokines and cell surface markers simultaneously, thus allowing detection of cytokines at the single cell level. A significant cost savings is provided, since there is no need to measure cytokines using separate ELISA techniques.

Other new procedures which can reduce the cost of performing assays use flow cytometry. For example, Hulstaert et al [24] have described a method to screen for HLA-B27 using flow cytometry. Certain manufacturers, e.g., Becton Dickinson Immunocytometry Systems of San Jose, CA, offer kits and the appropriate software to perform the analysis. With this procedure, sample preparation is easy and fast, and the analysis can be done automatically. This process eliminates the complexity of setting up a microcytotoxicity assay. Hence this technique may be useful in laboratories which are using flow cytometry for other applications, and wish to expand their test menu in a cost efficient way.

Another whole blood kit is available to measure neutrophil function, including phagocytosis and oxidative metabolism by flow cytometry (Orpegen, Heidelberg, Germany). With their Phagotest[R] system, one can quantitatively determine leucocyte phagocytosis in unfractionated human blood. Whole blood is incubated with fluoresceinisothiocyanate labeled microorganisms. Monocyte and granulocyte phagocytizing activity is measured by flow cytometry. This procedure offers several advantages over the more traditional labor intensive functional assays performed in the past. Since whole blood is used, there is no need to isolate the specific cell populations, and the adhering bacteria can be distinguished from the ingested bacteria.

In the Bursttest[R] system, oxidative metabolism is determined in stimulated leucocytes by using Dihydrorhodamin123 as the fluorogenic substrate. The assay, which uses whole blood, can be performed in less than 1.5 hours. Individual cells can be measured by flow cytometry, allowing results from a heterogeneous population of cells. It must be noted that as of this writing, these two kits are for research purposes only and not for *in vitro* or *in vivo* diagnostic use.

Cost and time saving changes have been made in other tests as well. Tests for the measurement of specific IgE antibody have become more automated and faster, with better quality control monitoring. For example, in one system in use in our laboratory (Pharmacia CAP System) the assay TAT has decreased from 3 days to 3 hours. The system is more automated and now uses a microtiter plate instead of plastic tubes.

With an automated system each step has more consistency and there are regular calibration, maintenance, and checks in the system. These features enable the user to troubleshoot problems much easier and faster. The new system eliminates user interpretation of acceptability of standards and controls as it has specific criteria for acceptable runs, and alerts the user of unacceptable results with flags. Since much of the data collected on controls accumulates within the system, it is very easily retrieved for examination at a later time. The old system had no storage capability and made quality control monitoring very difficult and time consuming.

Other changes in procedures result in the use of safer materials and reagents. For example, in gel electrophoresis procedures it may be more cost effective and safer to purchase pre-cast gels than to prepare one's own. Alpha-1-antitrypsin phenotyping performed in our lab requires a gel media to perform isoelectric focusing of the serum alpha-1-antitrypsin protein components. The gels, prepared fresh for each run, are composed of non-polymerized acrylamide which is a strong neurotoxin. In addition to the advantages of saving technologists' time and being safer, precast gels provide more consistent gels from run to run, and produce better quality results.

Agar double diffusion techniques are often employed in an immunology laboratory. In our own laboratory, serum precipitins against a wide range of environmental and occupational antigens have been used in the evaluation of patients with hypersensitivity pneumonitis.

Many changes to this procedure have been implemented over the years. However, we continue to prepare our own slides although commercially prepared slides are available. The cost of the labor involved in preparing each slide is very little and the cost of reagents is nominal. However, a major modification that accelerated TAT was to decrease the normal incubation time from two days at room temperature to one day at 37°C. Decolorization of the slides was performed in the past by using an alcohol solution. Now, with new types of stains, the destaining can be accomplished with distilled water in a shorter time. Overall labor and the use of reagents are reduced, safety is improved, and TAT is accelerated.

Many labs consider home-made or home-brew tests to be less costly than buying a test-kit from a manufacturer. Often times there are no alternatives if the test is not available from a commercial source. Certainly the cost of purchasing reagents and supplies separately can be cheaper. However, troubleshooting expenses, particularly labor costs, can be excessive with a home-made assay.

X. CONCLUSIONS

As one can see, many technological changes have taken place in the immunology lab over the years. Automation, new instrumentation, and enhanced proficiency have significantly modified the way tests are performed. New instrumentation has allowed the lab to analyze samples faster and more accurately with lower labor costs and increased productivity. Many changes have contributed to a safer working environment within the lab, and often result in alternatives which are more environmentally friendly.

Enhanced computer software has provided better documentation and monitoring of results and quality control, and has provided more efficient communication between laboratories. Computers have eliminated certain clerical errors by allowing transmission of information from one system to another without user interaction. This is of great benefit for acquiring fast, accurate results; ultimately costs are reduced and the quality of patient care is enhanced. Designing an adequate and efficient workspace can decrease safety risks, increase productivity, and allow for growing room.

Controlling costs of operating a clinical immunology laboratory can be a challenge. By standing back and reviewing the specific functions that the laboratory performs, with an eye towards efficiency and greater productivity, may result in savings. This will place your laboratory in a good position to survive in the ever changing competitive marketplace.

XI. REFERENCES

1. Bellanti, JA. Role of the clinical immunologist. 7th Annual Meeting, Association of Medical Laboratory Immunologists. Boston, MA, 1994.
2. Deodar, SD. Organization of the clinical immunology laboratory. In Manual of Clinical Laboratory Immunology, 3rd edition, N. Rose, H. Friedman and J. L. Fahey, editors. 1986. American Society for Microbiology, Washington, D.C. pp 948 - 950.
3. Perlmutter, LL: Role of the clinical immunology laboratory today. 7th Annual Meeting, Association of Medical Laboratory Immunologists. Boston, MA, 1994.
4. Cook, J: Cost-containment challenges met with workstation consolidation. Advance, 8-9, June 13, 1994.
5. Conn RB: Health care reform and the clinical laboratory. Clin Chem 39(9), 1759 - 1760, 1993.
6. Nakamura, RM and DJ Bylund: Factors influencing changes in the clinical immunology laboratory. Clin. Chem. 40:2193-2204, 1994.
7. Check, W: Curbing costs in your microbiology lab. CAP Today, 8:34, 1994.
8. Nace, L. Reconditioned equipment, viable option in today's financially strapped lab. Advance 6(33): 6, 1994.
9. Bodin, S. Immunoassay automation: saving time, labor and money. MLO. January, p. 34-36, 1995.
10. Clinical applications of flow cytometry: Quality assurance and immunophenotyping of peripheral blood lymphocytes. NCCLS Document H42-T, Vol. 12(6), 1993.
11. Harbeck, RJ and Giclas, PG: Manual of methods and procedures in clinical laboratory immunology. New York, Raven Press, 1991.
12. Buckley, RH: Primary immunodeficiency diseases. In: J. Wyngaarden and L. Smith, eds. Cecil Textbook of Medicine. WB Saunders Co., Philadelphia, Pa. 1446-1453, 1992.

13. Harbeck, R: Laboratory evaluation of immunodeficiencies. ASCP Spring 1993 Teleconference series.
14. Mroz, MM, Kreiss, K, Lezotte, DC, Campbell, PA, and Newman, LS: Reexamination of the blood lymphocyte transformation test in the diagnosis of chronic beryllium disease. J. Allergy Clin Immunol 88:54, 1991.
15. Harbeck, RJ, Hoffman, AA, Redecker, S, Biundo, T and Kurnick, J: Separation of human polymorphonuclear leukocytes and lymphocytes on Percoll gradients. Clin. Immunol. Immunopath. 23:682, 1982.
16. Rice, JE and LP Bignold: Chemotaxis of polymorphonuclear leukocytes in whole blood in the 'sparse-pore' polycarbonate (Nucleopore) membrane/Boyden chamber assay. J. Immunol. Methods 149:121, 1992.
17. Junger, WG, Cardoza, TA, Liu, FC, Hoyt, BB, and Goodwin, R: Improved rapid photometric assay for quantitative measurement of PMN migration. J. Immunol. Methods 160: 73, 1993.
18. Mosmann, T: Rapid colorimetric assay for cellular growth and survival:application to proliferation and cytotoxicity assays. J. Immunol. Methods 65:55, 1983.
19. Sladowski, D, Steer, SJ, Clothier RH, and Balls, M: An improved MTT assay. J. Immunol. Methods 157: 203, 1993.
20. Ahmed, SA, Gogal, RM, and Walsh, JE: A new rapid and simple non-radioactive assay to monitor and determine the proliferation of lymphocytes: an alternative to [^{3}H] thymidine incorporation assay. J. Immunol. Methods 170: 211, 1994.
21. Fletcher, MA, Klimas, N, Morgan R, and Gjerset, G: Lymphocyte proliferation. In Manual of Clinical Laboratory Immunology, 4 th edition, N. Rose, E. Conway de Macario, J. L. Fahey, H. Friedman and G. M. Penn, editors, 1992, American Society for Microbiology, Washington, D.C. 213-219.

22. Maino VC, Suni, MA, and JJ Ruitenbery. Rapid flow cytometric method for measuring lymphocyte subset activation. Submitted for publication, 1994.
23. Jung T, Schauer, U, Heusser, C, Neumann, C and Rieger, C: Detection of intracellular cytokines by flow cytometry. J. Immunol. Methods 159: 197, 1993.
24. Hulstaert, F, Albrecht, J, Hannet, I, Lancaster, P, Buchner, L, Kunz, J, Falkenrodt, A, Tongio, M, De Keyser, F, Veys, EM, Noens, N, Mir, N, Costello, C, Becker, R, Strauss, K: An optimized method for routine HLA-B27 screening using flow cytometry. Cytometry 18:21, 1994.

INDEX